The Sun

Claudio Vita-Finzi

The Sun

A User's Manual

 Springer

Claudio Vita-Finzi
Natural History Museum
London
UK

ISBN 978-1-4020-6680-5   e-ISBN 978-1-4020-6881-2
DOI: 10.1007/978-1-4020-6881-2

Library of Congress Control Number: 2008925139

Printed on acid-free paper

9 8 7 6 5 4 3 2 1

springer.com

For Penelope

Preface

Few of us have any idea of how the Sun works and how it affects our lives beyond the obvious business of night and day and summer and winter. Yet we cannot make sensible decisions about dark glasses or long-distance air travel or solar panels, or fully understand global warming or the aurora borealis or racial characteristics, without some grasp of the workings of our neighbouring star. And quite apart from questions such as these, many of us may just be curious. The 19th century American poet Walt Whitman became *tired and sick* in a lecture by *a learn'd astronomer* and wandered out, *in the mystical night air, to look up in perfect silence at the stars*. If he'd concentrated a bit harder in class he would have started noticing all sorts of new marvels in the sky.

The Sun is intended for the curious reader. Some of the material is hard but no more so than you find in a decent biography or a gardening manual. In any case the sticky bits can be skipped on first reading (or forever), although I suspect anyone who is really interested in the world outside the window will relish getting his or her mind around neutrinos, cosmic rays, and even a dash of relativity, and will not want to be patronized.

The book is designed to portray some of the myriad ways in which the Sun impinges on our lives. I had been working on a period of silting that affected the rivers of southern Europe and north Africa during the Middle Ages and that tends to be blamed on humans and their goats, and I found that it could be explained better and more simply by shifts in climatic belts caused by a flickering Sun. That led me to investigate how far the Sun's output does change over time and whether we can plan ahead to prepare for the next serious blip; and that in turn led to the early history of the Sun, its workings, and the many ways in which it interacts with humanity.

This brings me to my favourite moment on an Italian beach, when a fashion-conscious mother with one of those bandsaw Milanese voices called out to her little daughter 'Marisa, don't go in the water. You'll get your bathing suit wet.' What she should have said was 'if you stay in the August sun between 11 and 2 pm and get burnt three times you will increase the odds of getting skin cancer as an adult by 60% and even if you don't your face will look like a prune.' But no such simple formula for mothers is yet available, nor do I advocate that, like radiologists and nuclear power engineers, children should wear radiation badges. All I do is try to explain *how* we toast so that the reader can choose his or her sun lotion rationally.

But there is much more to the Sun than sunbathing, and I try to follow the same approach in discussing human evolution, climate change, solar energy, the Sun's effect on radio broadcasts, and the internal workings of the Sun itself. I do go on a bit about hydrogen and helium but my excuse is that they make up the bulk of the visible matter in the Universe. Similarly wavelengths, which, like frequency, can be used to describe the behaviour of different kinds of solar energy from X-rays to radio waves. You do not have to be a geek to appreciate such matters, witness a useful mnemonic for the relationship between wavelength and frequency to be found in one of the tales of diplomatic life by Lawrence Durrell:

"If there is anything worse than a soprano," said Antrobus judicially, as we walked down the Mall towards his club, "it is a mezzo-soprano. One shriek lower in the scale, perhaps, but with higher candle-power."

Just bear in mind that he got it the wrong way round.

There are many paradoxes in my account. The Sun drives the weather and keeps the Earth's temperature at tolerable levels, it is the basis of photosynthesis and thus the life of plants and the creatures they sustain, and its magnetic field shelters us from dangerous cosmic rays; yet at the same time the ultraviolet (UV) part of the solar spectrum may damage DNA and human tissue, solar flares can destroy spacecraft, power systems and computers, and there is every indication that the Sun precipitated a mini Ice Age less than two centuries ago. Sunshine allows us to generate vitamin D but too much of it can lead to skin cancer and cataracts. Etcetera etcetera.

As is by now obvious, and the end notes confirm, my sources range from astronomy to archaeology and from geology to genetics. The references are numerous, but it seems unjust not to give credit to the boffin who has slaved for years to bring you a vital piece of nature's mosaic, and you are free to ignore the tiny superscript numbers that lead to the fountainhead. There are many excellent books on each of the topics I discuss but so far as I know none that tries to cover all the topics at introductory level. Unfamiliar terms and abbreviations are defined when first used. Although astronomers normally employ the Kelvin temperature scale I have stuck to degrees Celsius (°C) as the book deals with everyday temperatures on Earth as well as those within the Sun's interior where -273.16°C (zero on the Kelvin scale) hardly makes a difference to 15,000,000 K. I use the power notation (10^{10}, for example, for 10,000,000,000) or Myr (for a million years) when a row of noughts, as you can see, is no more informative.

The following have done their generous best to weed out errors of fact on my part in the sections that do not deal with river mud: John Adams, Paul Bahn, Benedetta Brazzini, Charles Cockell, Eric Force, Ian Maddison, Ken Phillips and Ray Wolstencroft. I am also indebted to the late Rhodes Fairbridge for introducing me to Springer, to Petra van Steenbergen, Hermine Vloeman, Padmaja Sudhakher and Maury Solomon there for much support, to Don Braben, Annette Bradshaw, Ann Engel and Penelope Vita-Finzi for astringent comments on an early draft chapter, to Tony Allan, Geoff Bailey, Roger Bilham, Stephen Lintner and Ian Maddison for references, to Leo Vita-Finzi for matchless advice, to John Burgh and Rick Battarbee for musical solace, to Simon Tapper for help with the figures, to the many who generously supplied figures (and are acknowledged in the captions), and to the engineers and scientists responsible for the SOHO (Solar and Heliospheric Observatory) satellite, which was launched jointly by the European Space Agency and NASA in 1995 with a 'nominal' life of 2 years and is still busily at work as I write.

London, January 2008

Who … would not wish to know what degree of permanency we ought to ascribe to the lustre of our sun? Not only the stability of our climates, but the very existence of the whole animal and vegetable creation itself, is involved in the question.

John Herschel, *Treatise on Astronomy*, 1833

Contents

Chapter 1
Looking at the Sun

If curiosity, as Isaac Asimov has eloquently argued, is one of the noblest properties of the human mind, then prediction is its richest reward. And its survival value is obvious. Is the tide about to turn? Do we need more firewood? When will the herds come back?

Some of the best evidence for effective forecasting in prehistory comes from success in the hunt. In the Dordogne region of France, several of the late Pleistocene sites renowned for their rock art and flint work show great economic dependence on reindeer. At the Abri Pataud, for instance, reindeer make up between 85% and 99% of the bones left by its prehistoric occupants.[1] The caves and shelters open onto valleys bordered by steep cliffs which would have created natural corrals in which to confine reindeer transiting between their summer and winter grazing areas. To judge from the bones the cave occupants timed their seasonal visits shrewdly, even if the reindeer they caught did not. Although the first few seasons must have been a matter of trial and error it seems likely that hunting proficiency in the Palaeolithic came to depend a good deal on observing seasonal clues of one kind or another: the first thaw, for example, or the flowering of some dependable shrub, or the departure or return of a migrant bird.

Most prehistoric hunters and gatherers moved periodically to exploit food that was seasonally abundant. In Alaska they did so for berries, shellfish, deer, fish and sea mammals. To be sure, as in much of the panorama of natural selection, we rarely come across the failures: the luckless family which spent the winter forlornly looking for whelks did not leave massive shell middens behind. But there are countless heaps of food remains which reflect seasonal shrewdness and which imply at least a measure of planning.

Sun as clock

Success in such enterprises was more assured once a link was found with the stars, the Moon and the Sun, initially signalled, perhaps, by a change in the length of a distinctive shadow or the illumination of the blank canvas of a smooth rock face. At high northern latitudes the noonday Sun is at its highest in the sky in summer, retreats south to its furthest position in winter, and then gradually returns. Even in

C. Vita-Finzi, *The Sun: A User's Manual*,
doi: 10.1007/978-1-4020-6881-2_1, © Springer Science + Business Media B.V. 2008

the monotonous tropics plant life responds to what has been called the drumbeat of the solar year.[2]

Some of the most ancient human structures commemorate the solar year. At the Newgrange passage tomb in the Boyne valley of Ireland, dating from about 3000 BC, the sun at the midwinter solstice shines for a few minutes though the roof box and illuminates the back wall. The axis of the passage corresponds within about 5' to midwinter sunrise at the time the tomb was built.[3] What is perhaps the oldest solar observatory in the Americas, dating from the 4th century BC, was recently excavated at Chankillo in Peru. A series of 13 towers aligned north-south along a low ridge form a "toothed" horizon which, viewed from observation points to the east and west, allow the rising and setting positions of the Sun to be observed at intervals between the winter and summer solstices.[4]

The vast effort required to erect these monuments, where a few sticks would have done the job equally well if timekeeping is all that was required, shows that some kind of ritual accompanied, as it still does, the practical inauguration of a fresh set of seasons. To be sure, there is a strong temptation to read too much wisdom in such alignments. Take, for example, Stonehenge, the mighty complex of earthworks and standing stones built in at least seven stages between 3100 BC and 1900 BC on Salisbury Plain in southern England. The consensus is that Stonehenge was designed to mark the position of sunrise at the summer solstice. The question is whether, besides any religious and social ceremonial associated with that annual event, the stones and banks had any other astronomical function.

An elaborate analysis of Stonehenge and other stone monuments was published in 1909 by Norman Lockyer,[5] who concluded that Stonehenge was a solar temple, as indicated by the alignment of its 'avenue', which marked sunrise on the longest day of the year. This event had, as he put it, not only a religious function: it had also the economic value of marking officially the start of an annual period. But Lockyer did not rule out other 'capabilities' for Stonehenge, such as a connexion with the equinoxes or the winter solstice.

Lockyer used a theodolite, and pen and paper, to make his case. The advent of the computer made even more elaborate analyses possible, and in 1966 the American astronomer Gerald Hawkins presented evidence for Stonehenge as an ancient computer which, among other things, could be used to predict lunar eclipses. The astrophysicist Fred Hoyle went on to suggest in 1977 that Stonehenge was in effect a model of the Solar System and could be made to function as a computer which was even more precise than Hawkins had claimed as it could predict lunar eclipses to the day. There the matter rests, but uneasily, as archaeological excavation continues to reveal more traces of the alleged computer and the order in which it was assembled and repaired.

Much doubtless depended at these ancient observatories – if that is what they were – on shutters and markers of one sort or another which have long turned to dust. The remarkable success ancient Greek astronomers had in tracking and recording heavenly motions likewise appears remarkable partly because we have little trace of the devices with which they made and documented their observations. Consider the phenomenon of precession (strictly speaking the precession of the equinoxes), the cone-shaped path followed by the north Pole and, as we now know, completed in the space of 25,770 years. Hipparchus of Nicea (190–120 BC) had identified the effect in 150 BC or thereabouts by reference to observations made by his predecessors even though the movement amounts to about 1° per 72 years. That achievement argues for good eyesight (as there were no telescopes), stable instruments and dependable archives.

However, the 'Antikythera instrument', discovered in 1900 near Crete in a sunken cargo ship full of statues, suggests that we have underestimated the technology that underpinned the Greek achievement. The device was made of bronze, now badly corroded, and housed in a wooden case measuring about $33 \times 17 \times 9$ cm. Its main function, so far as one can tell from its gear wheels and fragmentary engraved inscriptions, and after a century of study combining the skills of computer scientists and historians of astronomy with the results of X-ray tomography, was to predict the position of the Sun and Moon and perhaps also the planets. Apparently the mechanism, which dates from 150–100 BC, even allowed for variations in the Moon's motion across the sky. It may have been based on heliocentric rather than the geocentric principles then prevailing, and it indicated position in the Saros cycle and a longer eclipse cycle. The Saros cycle, known to the Babylonians, is the period of 18 years and $11\frac{1}{C}$ days after which the Sun, Earth and Moon return to the same relative position in the heavens.[6]

The solar year

As at Stonehenge, the focus in Greece was on both the Sun and the Moon. The lunar cycle is not straightforwardly related to the solar year. The synodic cycle is the time it takes the Moon to complete a cycle of phases and occupies 29.53 days, so that 12 such cycles total 354.4 days and 13 cycles total 383.9 days. It is impossible to say when an attempt was first made to harmonise the solar and lunar years, but there is some evidence for a tally of lunar phases in Palaeolithic times. The American scholar Alexander Marshack found scratches and cuts on a piece of bone dating from an estimated 30,000 years ago in the Abri Blanchard, near Sergeac in the Dordogne region of France, which he thought represented the phases of the moon over 2¼ lunar months. The Taï bone plaque, dating from about 12,000 years ago, shows sets of 29 notches, which Marshack equated with the synodic month, the average time taken by the Moon to run through a complete cycle of phases.[7]

Whatever the validity of such claims, the lunar month was the basis of the calendar in many societies, including the Sumerians, the Babylonians and the ancient Greeks. Indeed, the lunar calendar has been retained by Muslims and Jews, and by Christians for their movable feasts. But impatience with the mismatch between the lunar calendar and the seasons in the end weakened and then eliminated the Moon's calendric preeminence in many cultures.

By 2000 BC the Sumerians had adopted a year of 12 months of 30 days. Some 1,500 years later the Babylonians squared their lunar calendar with the seasonal or solar cycle by allocating an extra month to 7 years out of every 19. The Greeks retained a lunar calendar but added 90 days to it every 8 years. The Jews added a month every 3 years supplemented from time to time by an additional month. The

Chinese calendar is a combined solar/lunar one for which records inscribed on oracle bones date back to the 14th century BC.[8]

In the Nile valley the solar and lunar calendars were harmonized as early as the fifth millennium by the addition of 5 days to the 360 of the lunar year. Later the start of the year came to be marked by the heliacal rising of the dog star Sirius, that is to say the time when it first became visible above the eastern horizon, but as this was found to occur 6 h later each year, an additional 1/4 day then had to be included as a leap day every four years. The need to safeguard the solar year was once again a key concern.

This was the calendar adopted by Julius Caesar and named Julian after him. At the Council of Nicaea in AD 325 the Emperor Constantine decided that Easter should fall on the first Sunday after the first full moon after the spring equinox according to the Julian calendar. In 1267 the friar Roger Bacon wrote to the Pope to warn him that the official date for the spring equinox was 9 days late. In Bacon's view any layman could tell this was the case by looking at the changing position of the sun's rays on his wall.

The Julian calendar remained in force in the West until the 16th century, by which time it was clear that 365¼ days was an overestimate (by 11 min and 14 s). The discrepancy was put right by Pope Gregory XIII, who decreed that the day following 4 October 1582 would be 15 October, and that 1700 and other end-of-century years would no longer be leap years unless divisible by 400. The Old Style (Julian) calendar was retained in countries not in thrall to the Pope: in England and its dominions, for example, until 1752; it still governs the Greek Orthodox Church. And for some astronomical tasks it is convenient to reckon the passage of time in Julian days, that is to say by the number of days that have elapsed since Greenwich mean noon on Monday 1 January 4713 BC. The Julian date (JD) then is the Julian day number (JDN) followed by the fraction of the day that has elapsed since the preceding noon. Thus the JD for Monday 7 January 2008 at 1800 hrs is 2454473.25.

For normal tasks we cleave to the Sun as yearly measure. Even Napoleon's Revolutionary Calendar began on the autumn equinox of 1792. (The calendar lasted only until 1 January 1806). The solar day changes in length throughout the year both because the Earth's orbit is elliptical, so that its rate of progress must vary, and also because the Earth's axis of rotation is tilted with respect to the Sun's path through the celestial sphere (the ecliptic). In this respect a sundial is superior to any mechanical (or chemical) clock because it faithfully indicates the interval between successive local noons. It can even be made to allow for the equation of time, as the variation in hour length during the year is called, by having curved rather than straight hour lines.

Sundials are doubtless the oldest timepieces. An example from Egypt dates from 1350 BC. The invention of the magnetic compass much benefited the use of portable sundials, which were made in pocket form well into the 19th century. The sundial could of course serve for navigation by being adjusted periodically for local time whereupon the shadow of the gnomon would allow the chosen bearing direction to be followed. A Viking sun compass which worked on this principle and dates from AD 1000 has been found in Greenland. During the Second World War the sun compass came into its own again in North Africa, when long distances had to be covered over featureless terrain under clear skies in vehicles whose moving metal parts reduced the accuracy of magnetic compasses. It proved highly compatible with the bubble sextant, which had been designed for navigation from aircraft to provide an artificial horizon as reference for measuring the elevation of the Sun or a star.[9]

We now use atomic clocks to correct for the unsteady progress of the Earth around the Sun and also to trace changes in the Earth's rotation, which is gradually slowing largely because of the braking effect of the tides. The second was formerly defined as 1/86,400 of a mean solar day and, once the day was found to be inconstant, as 1/86,400 of the mean solar day 1 January 1900. It is now defined as the duration of 9,192,631,770 cycles of radiation corresponding to the transition between two hyperfine levels of the ground state of caesium 133 (^{133}Cs). This new second is the time unit that underpins the management of GPS satellites. It also serves for distance measurement on Earth using signals from quasars far in Space in order to investigate such matters as the relative movements between the continents.[10] Even so a leap second is introduced in some years to keep the difference between international atomic time (TAI) and mean solar time to less than 0.9 s a year: the solar year rules.

Sun as god

How far progress in recording the motions of the Moon and the Sun was matched by improved understanding is not always clear. In many societies astronomy was inseparable from religion, divination and a centralized authority, and it was doubtless politic to retain its symbolic trappings. The Babylonian sun god Shamas, for example, would emerge from a vast door on the horizon every morning, mount his chariot and cross the sky to the western horizon, where he entered another door and travelled through the Earth until he reached his original starting place by the next morning.

But perhaps the error lies in equating vivid imagery with ignorance. Many terms in physics, for example, employ analogies or homely terms which may mislead more than they explain. The spin of atoms, protons or electrons, for instance, though associated with angular momentum and with magnetic moment, is not rotation in the sense of classical mechanics. In particle physics flavour, charm, topness and strangeness are categories proposed by the physicist Murray Gell-Mann which were intentionally whimsical, just as a quark, three of which make up a baryon (baryons include protons and neutrons), alludes to Three quarks for Musther Mark in James Joyce's Finnegan's Wake. This is not to suggest that the Babylonians were a particularly whimsical people but that, as with present-day religions, the celebrants were surely able to juggle imagery with commonsense. Fig. 1.1

The imagery on occasion actually proved a convenient device for correcting the current calendar. Nut, the mother of all Egyptian gods, accounted for the daily solar cycle by swallowing the Sun every evening and giving birth to it every morning in the shape of the scarab beetle, Khepri (Fig. 1.1). The Sun god Ra would then ride west in his sacred boat across the sky until sunset, where he was swallowed again. When it became clear that the length of the solar year needed adjusting the correction was blamed on her gynaecological problems, as she required an extra 5 days to bring several pregnancies to term. Whether borrowed or dreamed up afresh the metaphor of a radiant object crossing the sky in some kind of vehicle recurs in succeeding centuries. In Bronze Age Europe the sun traverses the sky in a chariot.

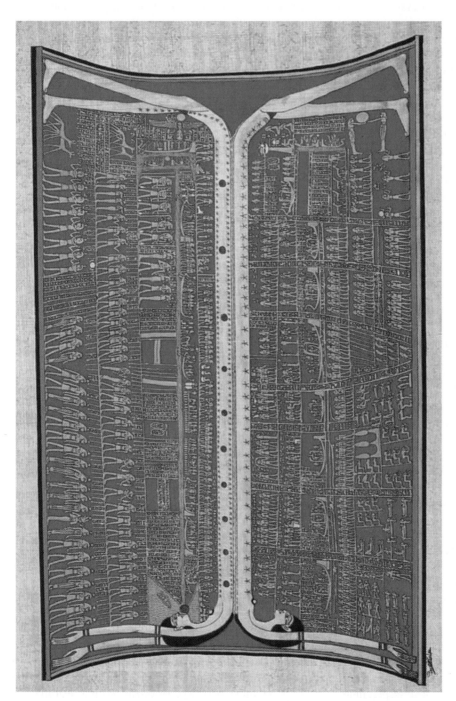

Fig. 1.1 The Goddess Nut swallows the Sun at dusk and gives birth to it at dawn. Painted ceiling of the tomb of Rameses VI (20th Dynasty, 12th century BC). The image of Nut representing the Book of the Day displays 10 solar disks along her body, one in her mouth and one being born bearing the image of the dung-beetle Khepri, symbol of rebirth (Courtesy of Anthony Kosky, Copyright 1991)

In Greek mythology Helios was imagined as a god crowned with the solar halo who drove a chariot across the sky each day and night.

It was a short and tempting step for the human ruler to identify with that shining figure. The archaeologist Jacquetta Hawkes[11] argued that, once the pattern of movement among the Sun, Moon and planets had been to some extent comprehended, the Sun God was accepted as its master, and the earthly ruler in Mesopotamia, Egypt, Mexico or Japan came to be seen as its agent or even its incarnation. Pharaoh Amenhotep III, for example, was 'the dazzling sun'. Atahualpa, killed by Pizarro, was the last of the Inca sun-gods. The Persian kings ruled by divine grace and accordingly received a fiery aureole as a gift from the Sun God. Gold was the chosen substance and a wheel the symbol. The god sometimes demanded a price for defeating darkness. For the Aztecs the Sun's arrival each day could be guaranteed only by the regular sacrifice of pulsating human hearts (Fig. 1.2). The arrangement seemed to work.

Regeneration, recurrence, periodicity and the struggle between light and dark are common themes in solar mythology. The cult of the Unconquered Sun, introduced by the Roman Emperor Aurelian in AD 274 and celebrated on 25 December, is perpetuated in the art and ceremonial of the Christians. In Peru, Garcilaso de la Vega[12] reported in the *Comentarios Reales de los Incas* in 1609

> Of the four festivals which the Inca kings celebrated in the city of Cuzco, which was another Rome, the most solemn was the festival of the Sun in the month of June, which they called Inti Raimi, meaning the solemn resurrection of the Sun. They … celebrated it when the solstice of June happened.

Eclipse as weapon

Not all representations or modes of veneration of the solar deity embodied profound astronomical truths: the wheel could denote movement, or the fiery disk, or neither. But eclipses would surely prove a useful device for cowing the multitudes.

Columbus used a lunar eclipse in 1504 to impress an Amerindian community with his powers when they threatened to cut off his supplies. Lunar eclipses are relatively easy to forecast and can be viewed from anywhere on the night side of the Earth. During a solar eclipse, however, the Moon's shadow on Earth is at most 270 km wide and its path is both narrow and difficult to predict without a very precise knowledge of the Moon's orbit (Fig. 1.3).

There is, moreover, no simple pattern of recurrence. The first known report of an eclipse of the Sun was made in China in 2136 BC although the oldest true record was made in 1375 BC at Ugarit in Mesopotamia. Prediction was apparently delayed until the 1st century BC and even then was based not on a full grasp of the orbital complexities but on the Saros cycle. Cuneiform experts claim that the Babylonian astronomers could predict solar as well as lunar eclipses as early as the 4th century BC. Thus Tablets BM 36761 and 36390 predict a solar eclipse for 6 October 331 BC; the translators remark 'As a matter of fact a solar eclipse did take place … but it could be watched in Greenland and North America, not Babylonia'.[13]

Once solar eclipses could be predicted the scope for playing on gullibility blossomed. In Mark Twain's *A Connecticut Yankee in King Arthur's Court* the hero in

Fig. 1.2 Aztec sacrifice as nourishment for the Sun god Huitzilopochtli to ensure the Sun's daily journey across the sky (Courtesy of Prof. G. Santos)

AD 528 is about to be burnt at the stake but he secures his release by predicting a solar eclipse. So does Hergé's Tintin in *Prisoners of the Sun*, with the Incas unfairly portrayed as astronomically inept.

   In Babylon the gods used heavenly signs as warnings, and the astronomers meshed their observation with earthly events, such as the level of the River Euphrates or the price of barley, to construct the Astronomical Diaries (now in the British Museum in London) and thence to devise omens. The cuneiform tablets in question range from the 8th to the 1st century BC. Eclipses warn of imminent danger. A solar eclipse on 29 Nisannu (12 May), for example, meant that the king would die within a year. Alexander was accordingly warned by the Babylonian astronomer Bêl-apla-iddin[14] to avoid Babylon and appease Marduk, the supreme god of Babylonia, by rebuilding his ziggurat. Alexander agreed but then changed his mind, entered Babylon, and on 11 June he died.

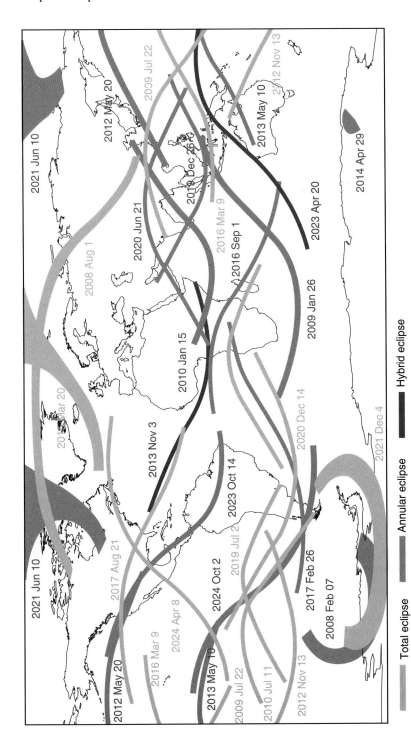

Fig. 1.3 Predicted solar eclipse paths (Courtesy of NASA at sunearth.gsfc.nasa.gov/eclipse/SEatlas/SEatlas3/SEatlas2001.GIF). Note that the paths are irregular and of very variable width. In a total eclipse the Moon totally blocks radiation from the Sun's photosphere; an annular eclipse occurs when the Moon covers only the central part of the Sun; a hybrid eclipse is seen as a total eclipse in some parts of the Earth and as an annular eclipse in others

The Sun as astronomical object

Astrology apart, the Greeks profited from the many centuries of detailed observation made in Sumer and Babylon both in data, such as an improved estimate for the length of the year, and in astronomical procedure, but much of their achievement can only be put down to native genius. By the 2nd century BC the Greeks were using the Sun's elevation to calculate the Earth's diameter and from there the distance to the Moon and to the Sun. Archimedes reports that among those responsible for these remarkable feats, Aristarchus of Samos (about 310–230 BC) believed that the Earth went round the Sun. It is said he made few converts because he could not prove what he claimed and in any case the suggestion was considered impious. The Earth was to stay in the centre of the Universe until after 1543, when Copernicus published *The Revolution of the Heavenly Bodies*, even though Arab scholars had transcribed Greek astronomical references to the Sun during the Middle Ages.

The Sun-centred model of the solar system is usually presented as the key item in the dispute between Galileo and the Church in order to underline the Church's ignorant intransigence. Galileo had offended the authorities not only by espousing the Copernican model but also by showing that the Moon departed from the Aristotelian ideal in having mountains and valleys, and that Venus also went round the Sun as it had phases just like our Moon. But from our present viewpoint his key contribution was to make the Sun a reasonable subject of scientific study rather than the object of uncritical veneration.

He also helped to make it safe for astronomers. Before the telescope was introduced (in about 1605), observation of the surface of the Sun often relied on various natural filters to protect the naked eye, such as thin haze and dust. In China certain kinds of jade were used for this purpose.[15] The story goes that Galileo went blind because he gazed at the Sun through his newfangled telescope, but (as a quotation below shows) he was well aware of the risks of sungazing.

Sunspots

The telescope, which Galileo perfected from a Dutch spyglass, boosted observation and also risk. In 1610 Thomas Harriot could train his x10 telescope on the Sun only soon after sunrise or before sunset if there was mist or thin cloud and even then for a minute or so at most.[16] Harriot was the first to record sunspots, which are marks on the Sun's disk, singly 100–100,000 km in diameter and in groups spanning up to 150,000 km. Galileo Galilei is sometimes credited with their discovery. In fact there is at least one Chinese report of a sunspot dating from the 8th century BC or perhaps even earlier. Theophrastus mentioned sunspots in the 4th century BC, and there are accounts of single sunspots from the 9th century AD in Europe and the 10th century in Arabia. The oldest known drawing of a sunspot dates from 8 December 1182 and shows the Sun with two black dots which are encircled by brown and red rings,[17] conceivably representing the dark central umbra surrounded by a brighter penumbra crossed by the bright radial structures of the typical sunspot. Galileo was probably not even the first to train a telescope on the spots, and has to share the glory with at least four others: David Fabricius (the Latinised version of Goldsmid) and his son Johann in Holland, Christopher Scheiner in Germany, and Harriot in England.

The idea of projection (Fig. 1.4), which came from Galileo's student Benedetto Castelli, brought with it several advantages: sunspots and other features were recorded during observation, rather than from memory; several observers could examine the same image simultaneously; and one could observe small spots which, in Galileo's words, were hardly perceived through the telescope and then only 'with great pain and damage to vision'. Astronomy has always progressed by recording, as seen in the ability of Hipparchus to use observations made in the preceding 150 years in his work on precession. The telescope made it possible for the argument in a factual work such as Galileo's sunspot book, *Istoria e Dimostrazioni Intorno alle Macchie Solari e Loro Accidenti* (1613), to be carried almost entirely by the illustrations (Fig. 1.5).[18]

The telescope also revealed, though in piecemeal fashion, that the Sun was not a smooth disk on which a few dark patches appeared from time to time: it had a complex morphology. In 1774 Alexander Wilson noted that sunspots appeared concave when viewed near the edge of the solar disk. This has been hailed as the first physical investigation of a sunspot and indeed the last until the 20th century, for his contemporaries and successors continued to focus on the number and distribution of the spots[19] even though Wilson, like William Herschel, the discoverer of Uranus, concluded that sunspots were holes through which the cool dark surface of the Sun could be glimpsed,[20] a notion which was long sustained by the belief that the Sun's composition was much like the Earth's.

The task of recording the Sun's changing moods was obviously much helped by photography. The first photograph of a sunspot was a daguerrotype taken in 1845. As imaging technology progressed, the apparently featureless surface between sunspots was revealed to be full of smaller spots no more than 180 km across.[21]

Colour coding

The question remained: what is the Sun made of? In 1835 the philosopher Auguste Comte declared that we would never be able to study the chemistry or mineralogy of a celestial object, or of any organic beings living on them. The development of spectroscopy in 1857 shows that *never* is a short time in science.

Light is made up of a spectrum of colours, as in the rainbow, and different substances when heated strongly give out light of a characteristic colour. For example, common salt, sodium chloride, when dropped onto a candle flame produces the yellow colour associated with sodium. In addition colour is a measure of temperature, as with a red hot or white hot poker. The spectroscope exploits these facts to distinguish between different elements and temperatures in the laboratory or through a telescope.

The different colours along the visible spectrum vibrate at different wavelengths, and the lengths are usually expressed in nanometres or billionths of a metre. As noted in the Preface, and as intuition tells you, the shorter wavelengths vibrate more rapidly and energetically: think of the waves that form when you shake a rope that is tied at one end or indeed the motion of different strings on a piano or guitar. The Sun emits energy over a great range of wavelengths – the electromagnetic spectrum (Fig. 1.6) –

Fig. 1.4 The projection method for observing the Sun: Christopher Scheiner and a fellow Jesuit scientist trace sunspots in Italy in about 1625 (Courtesy history.nasa.gov/SP-402/p9a.jpg). The caption shows that the bright areas known as faculae as well as sunspots (maculae) were already recognized

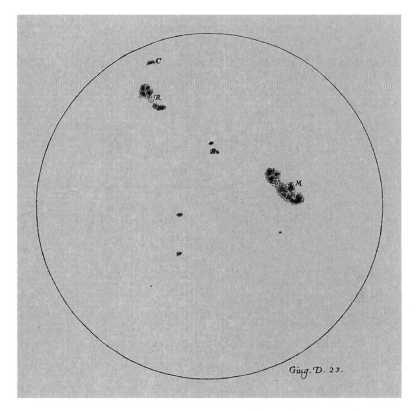

Fig. 1.5 Sunspots recorded by Galileo on 23 June 1612. The central umbrae and fringing penumbrae are clearly shown

of which visible light is a small portion. Newton had shown how a prism could be used to reveal the spectrum of visible colours making up white light. In about 1800 it was found that a thermometer just beyond the red end of the visible spectrum was affected by an invisible form of radiation, now termed infrared (IR), which we sense with our skins as heat. The scale was gradually extended in both directions to encompass the very long wavelengths of radio and the very short wavelengths of X-rays. The spectrum of light from the Sun (Fig. 1.7) also displays bright emission lines and dark absorption lines marking wavelengths where different elements absorb or emit light. Work of this subtlety requires the spectrum to be measured in ångström (Å), 1/10 of a nanometer or one ten billionth of a metre (1×10^{-10} m).

The Italian Jesuit Angelo Secchi used the spectrometer in the 1860s to classify 4,000 stars into four types, predominantly on the basis of their colour as a guide to their temperature. His categories were white and blue, yellow, orange, and red. A fifth class was later distinguished solely on the basis of spectral emission lines. By the end of the century Secchi's classification had been replaced by what came to be known as the Harvard scheme, which hinges on the strength of one of the hydrogen lines, but the essence of Secchi's scheme

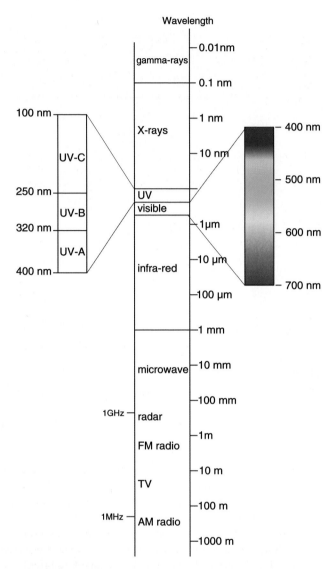

Fig. 1.6 The electromagnetic radiation (EMR) spectrum. Wavelengths (in metric units) and, for radio and TV, the corresponding frequencies (in Hertz). Note position of visible and ultraviolet (UV) bands. The UV subdivisions vary between authorities according to application. Medical sources sometimes favour 315–400 nm for UV-A, 280–315 nm for UV-B and 180–280 nm for UV-C

survives in the progression from blue (30,000–60,000°C) to orange red (2,000–3,500°C).

The total radiation emitted by one particular star, our Sun, and received by the Earth outside the atmosphere is known as the Solar Constant. It remained difficult to measure accurately primarily because the air is always in motion. The earliest attempts were

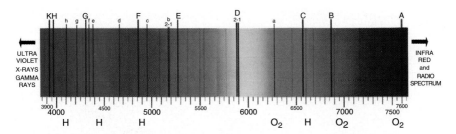

Fig. 1.7 Part of the solar spectrum showing lines (named after Joseph von Fraunhofer) which allow elements in the Sun to be identified from Earth. Lines for hydrogen (H) and oxygen (O₂) are indicated (After www.harmsy.freeuk.com, courtesy of Andrew Harmsworth)

made in the 19th century. In France C.S.M. Pouillet tried to allow for losses in the atmosphere and in 1837 came up with the figure of 1.8 cal/cm²/min. In 1881 a new device called a bolometer gave a figure of 3 cal/cm²/min on Mt Whitney at an elevation of 4,420 m. The result was more in error than the earlier estimate, to judge from the present accepted value of 1.94 cal/cm²/min, but the attempt had the benefit of revealing that atmospheric absorption was most pronounced in the UV part of the spectrum.[22]

The UV wavelengths are usually divided into UV-A (320–400 nm), UV-B (250–320 nm) and UV-C (100–280 nm). Ionizing radiation, a term which is applied to particles which are energetic enough to break down atoms or molecules into electrically charged atoms or radicals and which includes X-rays and the adjoining parts of the UV spectrum, forms a very small part of the solar spectrum. The atmosphere attenuates solar radiation in two main ways: by scattering and by absorption. Scattering is the work of air molecules, water vapour and aerosols, small particles that are suspended in the air such as salt or soot. Absorption is due mainly to ozone and water vapour, which absorb 97–99% of UV radiation at wavelengths between 270 and 320 nm.

The atmosphere is often cloudy as well as turbulent and climatologists welcomed the opportunity presented by balloon ascent, then rockets, and finally satellites to refine their data. The solar constant has been monitored by satellites since 1978 and found to average 1,368 watts per square metre (W/m²) at 1 astronomical unit (AU), the average distance between Sun and Earth or about 150 million kilometres. Half of the radiation is in the visible part of the electromagnetic spectrum and the remainder mainly in the IR part. The UV portion is minor but as we shall see of great significance to human wellbeing.

Besides temperature and composition the spectrometer proved capable of detecting motion towards or away from the observer as this leads respectively to a shift towards the violet or the red in the spectral lines. The redshift of light from distant galaxies is of course a key piece of evidence for an expanding universe. In accordance with the Doppler effect, which is familiar to us from the change in pitch of an approaching and receding siren, wavelengths are compressed as the source approaches and lengthened as it recedes, with a corresponding rise and fall in frequency. The effect allows us to measure the rotation of the Sun and the relative horizontal movement of parts of its surface over timescales measured in hours.

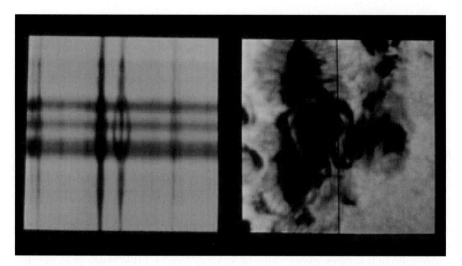

Fig. 1.8 Detecting the magnetism of sunspots by the Zeeman Effect. A single spectral line (i.e. a single wavelength) at the position shown by the line crossing the sunspot (right) is split by the magnetic field into two or three lines (left): a further example of how the properties of the Sun can be studied from Earth (after Lang 2006, courtesy of Kenneth R. Lang)

That is not all. The spectroscope demonstrated that the Sun, like the Earth, had a magnetic field with two poles analogous to the Earth's and, like it, running roughly over the poles (poloidal), and that parts of the Sun had their own subordinate fields. If the spectroscope scans across a magnetically active area, a single spectral line is split into two or three strands, and the width of the split is proportional to the strength of the field. The effect, named after its discoverer Pieter Zeeman (Fig. 1.8), showed that sunspots are magnetic and, by allowing changes in the field to be traced over time, it provides valuable clues to the workings of the solar interior.

A layered Sun

All these sources, in combination with centuries of earthbound observation, led to the recognition of three superposed, outer zones: the photosphere, which is the outermost visible part of the Sun, the chromosphere, which can be seen with the naked eye only during a total solar eclipse, and the corona, which merges with the solar wind. Though now detected by instruments on spacecraft, the solar wind was a revolutionary notion when it was first proposed (by E.N. Parker in 1958) as space was previously viewed as essentially a vacuum. Its existence was betrayed by the fact that comet tails point away from the Sun and by the occurrence of polar auroras at times when the number of sunspots suggested that the Sun was especially active. The flamboyant corona may well have inspired the solar wheels and haloes that accompany many of the sun Gods and that are especially prominent in Christian iconography.[23]

These remarkable theoretical and instrumental advances were being made despite the constraints imposed by the Earth's unstable atmosphere. The invention of the hot air balloon and then (appropriately) the helium balloon greatly reduced the problem. But balloons are unstable, and high-altitude observatories were the obvious answer. The Lick observatory, built in 1888, was at an elevation of 1,283 m; the Hale, on Mt Palomar, in 1946, at 1,710 m; the Keck, in 1992, at 4,200 m. The rollcall of dedicated solar observatories at high altitudes now includes the Mees Solar Observatory at the Haleakala ('House of the Sun') site in Hawaii (3,054 m), the Mauna Loa Solar Observatory (3,397m), the Teide Observatory (Tenerife) (2,400 m) and the Roque de los Muchachos Observatory (La Palma) (2,372 m) in the Canaries, and the Mt Evans Meyer-Womble Observatory in Colorado (4,312 m). The Indian Observatory at Hanle (4,500 m) is the highest but deals with optical and infra-red rather than solar astronomy.

But it was the development of rocketry, most infamously the V2, which transformed solar science. The Second World War had stimulated research into forecasting radio conditions and the interruptions caused by solar flares, although it would appear that some scientists emphasised military needs in order to promote solar physics pure and simple. This may apply to Wernher von Braun, who led the team responsible for designing the V2 ballistic missile that was used by the Nazis in 1944–5 against targets in Europe, as he had long championed space exploration and planned to use a V-2 to investigate the upper atmosphere in 1945. When the War ended, von Braun and 500 of his engineers moved to the USA, a move which culminated in the creation of the Saturn V launch vehicle and the successful Apollo programme.

In 1946 a V2 bore a spectrograph to record solar UV wavelengths, which until then had been obscured by the atmosphere. The period until 1957 saw many other successful flights. According to some commentators, the use of rockets did not in itself transform solar physics, even if it promoted activity in branches of the subject that were to bring great results. In their view the employment of radio telescopes to investigate the Sun likewise did not prove revolutionary. Similarly, the Soviet launch of Sputnik in 1957 sparked a general growth of capabilities for solar observation using ground-based telescopes, stratospheric balloons, high altitude aircraft, and rockets as well as spacecraft: at first satellites were seen primarily as complements to rockets, and solar observation lay third on the list of priorities.

It hardly needs saying that satellites and orbiters are now critical for solar research. They allow us to view selectively different parts of the Sun. and to monitor its changing complexion especially by extending the limited 400–700 nm range of human vision. The photosphere is usually viewed in visible light (Fig. 1.9a), but different wavelengths reveal loops, streamers and plumes that extend hundreds of thousands of kilometers above the photosphere. Prominences or filaments, for example, are well displayed in one of the hydrogen wavelengths known as the hydrogen alpha (Hα) line (656.3 nm). The use of a spectrometer adapted for solar studies, the spectroheliograph, makes it possible to see the chromosphere (Fig. 1.9b) without having to wait for an eclipse, notably in light at Hα or Ca II (393.4 nm) wavelengths, revealing the distinctive spicules, streamers of gas which can rise up to 15,000 km above the Sun. The corona, though briefly but flamboyantly revealed in

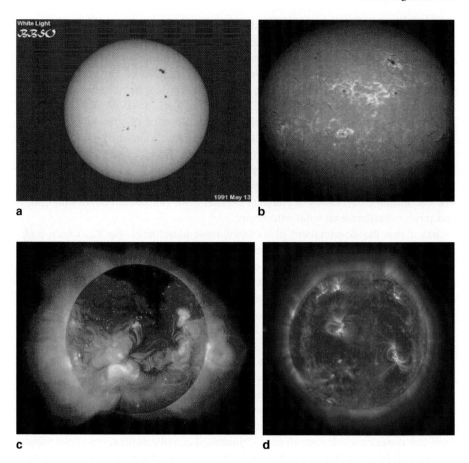

Fig. 1.9 Different faces of the Sun. (a) the photosphere imaged in white light in May 1991 showing sunspots; (b) the chromosphere picked out by H alpha showing filaments, dense cooler, clouds of material that are suspended above the solar surface by loops of magnetic field and plages, and bright patches surrounding sunspots; (c) the corona viewed in X-rays; (d) Sun'a magnetic field traced by loops of plasma picked out by UV (All images courtesy of NASA; the X-ray image was taken in 1992 with the Yokhoh Soft X-ray Telescope (SXT))

visible light during total eclipses, is readily captured at all times in X-rays (Fig. 1.9c) because of its very high temperature, but as the atmosphere is opaque to these wavelengths the imaging has to be carried out from spacecraft, such as SOHO and Yohkoh. The Sun's magnetic field can be picked out by loops of very hot plasma by UV imaging (Fig. 1.9d).

The construction of new observatories and the upgrading of existing ones reflect the continuing development of techniques for observing the Sun which are driven by curiosity as well as by the practical demands of the space age. For example, two international groups have responded to the demands of helioseismology, the use of waves (broadly analogous to seismology on Earth) to explore the structure of the Sun's interior. Both groups ensure continuous observation by their geographical spread. The Global Oscillation Network Group (GONG) consists of a network of six stations where sensitive and stable detectors can record almost continuously pulsations of the Sun's surface with a period of about 5 min. The observatories are in California, Australia, Hawaii, Tenerife, India and Chile. The BISON programme (Birmingham solar oscillation network), which is run from the UK, specializes in the analysis of data bearing on the Sun's core. It includes observatories in California, Chile, Tenerife, South Africa, Western Australia and New South Wales.

There is in this enterprise an echo of early attempts to measure the distance to the Sun by simultaneous measurement of the transit of Venus at two distant locations on Earth, a method promoted among others by Edmond Halley and adopted in 1761, 1769, 1874 and 1882. Even before that the estimate of the Earth's diameter by Eratosthenes in the 3rd century BC hinged on comparing the Sun's elevation at noon on midsummer's day in Alexandria and at Syene (Aswan), 5000 stadia to the south. (According to which conversion factor for a stadium we accept, Eratosthenes was spot on or 15% out.)

Satellites

The number and range of satellites and probes continues to grow in a vigorous effort to solve some long-standing puzzles about the Sun and its links with the Earth. The story begins between 1965 and 1968, with the four Pioneer series solar-orbiting, solar-cell and battery-powered satellites designed to obtain measurements of cosmic rays, the interplanetary magnetic field and other phenomena from widely separated points in space. (Note the belts-and-braces power source.) The Helios probes, launched in 1974 and 1975, had an even wider remit including measurements of gamma and X-rays and micrometeoroids. They set the record for closest approach to the Sun, at 0.3 AU, inside the orbit of Mercury, and continued to send data until 1985. The data from many deep space spacecraft were stored using a single coordinate system so that they could be readily compared.

There followed several missions partly or wholly devoted to solar and related issues. Explorer 49 (1973), for example, was placed in lunar orbit but devoted to solar physics. The Solar Maximum Mission (SMM) satellite, launched in 1980, was designed to investigate solar flares and solar radiation at many wavelengths. Not long after launch it suffered a power failure, but in 1984 the crew of the Challenger Space Shuttle were able to retrieve and repair the satellite, and it stayed in action until 1989. Later solar monitoring missions include the Upper Atmosphere Research Satellite (UARS, 1991, devoted to UV solar radiation) and ACRIMSAT (1999: total irradiance). The Active Cavity Radiometer Irradiance Monitor (ACRIM) instrument on SMM was the first to demonstrate that the solar constant was not constant. Even more fastidious are the measurements planned for Columbus, a laboratory destined for the International Space Station which will measure the Sun's output at wavelengths between 17 nm and 100 μm, thus encompassing 99% of the Sun's energy output.

Ulysses (1990) was designed to study the solar poles and interplanetary space above and below them. It used the gravitational pull of Jupiter to leave the ecliptic and passed the Sun's South Pole in 1994, 2000–1 and 2006–7 and the North Pole in 1995, 2001, and 2007–8 (Fig. 1.10). Yohkoh (1991) investigated flares and coronal disturbances with a telescope dedicated to imaging based on X-rays (2–10 nm).

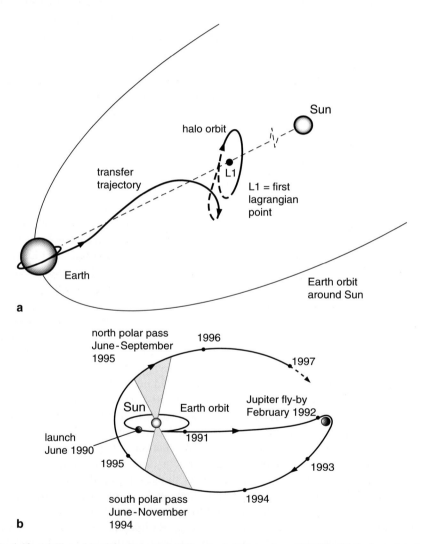

Fig. 1.10 (a) The orbit of the Solar and Heliospheric Observatory (SOHO); (b) the launch trajectory and orbit of Ulysses. The Sun's South Pole was visited again in November 2006–April 2007 and the North Pole between November 2007 and March 2008

The first space laboratory was Skylab, launched in 1973, an orbiting craft designed in particular for telescopic study of the Sun at various wavelengths and also to test human response to prolonged periods in space. As shown in Chapter 7 its career was cut short when a solar storm increased atmospheric drag to the point where the spacecraft fell to Earth but it was fruitfully occupied for 172 days and revealed the extent and violence of Coronal Mass Ejections (CMEs), which are massive eruptions of magnetized gas containing as much as 10^{13} kg of matter. The Solar and Heliospheric Observatory (SOHO) satellite takes advantage of the relative stability provided by the Lagrange point, a location where the gravitational pull between two bodies is in balance such as L1, the Lagrange point between Earth and Sun, but as the Sun is a powerful radio source which would swamp any signals from that point, SOHO, besides orbiting the Sun at the same rate as the Earth, follows a halo orbit around the Lagrange point, with each halo orbit taking 178 days. SOHO was launched in 1995 with a projected life of 2 years but its career was extended first to 2003, then to 2007 and perhaps finally to 2009.

The solar wind is implicitly one of the targets of most solar missions for what it says about the Sun's current behaviour and about the composition of the whirling cloud of dust and gas – the solar nebula – from which the Sun and planets are thought to have formed. Three missions have recently targeted it specifically. WIND, launched in 1994, spent most of its 10-year life time flying into the solar wind on the sunward side of Earth to investigate its physics and chemistry. GEOTAIL (1992) was a complementary mission focusing on the interaction of solar wind and Earth but downstream from the Earth. GENESIS (2001) attempted to collect a sample of solar wind on a set of delicate plates made of silicon, aluminum, gold/platinum, diamond and germanium. On 9 September 2004 it was due to be netted by two helicopters but its parachute failed to open and it crashed in the Utah desert. Even so some material – about 0.4 mg – survived for analysis.

Solar-B (2006) is designed to measure the Sun's magnetic field and thus help our understanding of violent solar events. The composition of the corona and of intervening interplanetary space is the target of the Advanced Composition Explorer (ACE) satellite launched in 1997 and in halo orbit at H1. Solar B also serves as space weather station and can provide a 1-h advance warning of any geomagnetic storms created by events on the Sun. The Solar Terrestrial Relations Observatory (STEREO), launched in 2006, goes one better because its two identical spacecraft, with one spacecraft ahead of the Earth in its orbit and one behind, provide a stereoscopic view of the Solar surface.

Many more solar missions are planned. They include the Solar Orbiter, which will perform close observations of the polar regions of the Sun, a difficult task from the Earth, at distances as short as 45 solar radii or $\frac{1}{5}$ AU. It is predicted that the images obtained will be up to ten times as sharp as any that can be taken now. Fifty years after it was first proposed, the Solar Probe will fly even closer – to within three solar radii – in order to investigate how the corona is heated and the solar wind generated.

Proxies and aliases

However safe and technically advanced it may be, a glimpse of the Sun will inevitably reveal only what it is now spewing out in radiation, energetic particles and magnetism. There remain other clues to the solar interior and its behaviour which are being ferreted out of some intriguing sources.

Perhaps the most potentially far-reaching is the hunt for the neutrinos which, as shown in the next chapter, are sub-atomic particles produced during fusion in the solar interior. The Sun's output of neutrinos is estimated to be 2×10^{38}/s, that is to say 200 million million million million million million. Exploding stars (supernovae) may produce in a flash 1,000 times the Sun's output during its entire life (ie in 9 billion years); and other neutrino sources include the events that attended the creation of the Universe. Yet only a handful of the particles will reach the Earth.

Large, subterranean detectors are required to trap and count these few arrivals. The first attempt to do this, in 1964, used a tank holding 45,000 l of purified perchloroethylene, a solvent used in dry cleaning, to detect the particles because an isotope of argon (^{37}Ar) which is radioactive is sometimes produced when neutrinos interact with chlorine in the solvent. The detector was placed 1,590 m below ground in a gold mine in South Dakota to exclude neutrinos produced by cosmic rays.

The experiment was a delicate one but the prospect of looking into the interior of a star pretty irresistible especially as the neutrinos in question would also provide information on the temperature deep inside the Sun. Attempts were later made in Russia and Italy using 15 tons of gallium. As neutrinos react with water to produce light which can be detected by sensitive instruments, the hunt was continued in the Japanese Alps 1000 m below ground using 68 tons of water and later heavy water and in Sudbury, Ontario (Fig. 1.11), at a depth of 2,000 m using 1,000 tons of heavy water[24].

Even more tangential as guides to the Sun's behaviour are the cosmogenic isotopes, such as carbon 14 (^{14}C) and beryllium 10 (^{10}Be), that are produced when galactic cosmic rays (GCRs) interact with the Earth's atmosphere and that accumulate in ice sheets, ocean sediments and tree rings. When GCRs – high-energy particles flowing into our Solar System from elsewhere in the galaxy or even outside it – reach the Earth's atmosphere they strike atoms and molecules to produce secondary particles. Some of the particles then react with nitrogen to produce carbon-14 and with oxygen, nitrogen and argon atoms to produce beryllium 10.

Incoming GCRs are deflected by the solar wind and the proportion that gets through thus provides an inverted measure of the Sun's level of activity. The Sun also emits particles which qualify as cosmic rays but these solar cosmic rays (SCRs) are relatively weak and when the solar wind is at its strongest any related increase in SCRs it may create is not sufficient to counteract the reduction in GCRs. Unfortunately – from the scientific point of view; in every other respect very fortunately – the Earth is also protected from cosmic rays by its magnetic field, and its variable effect complicates any attempt to evaluate the strength of the solar wind from the record of cosmogenic isotopes.

Fig. 1.11 External view if the detector of the Sudbury Neutrino Observatory, designed to capture solar neutrinos (Courtesy of Lawrence Berkeley National Laboratory). It began gathering data in 1999 and was designed to detect all three flavours of neutrinos

The effort entailed in coring the ice is easily justified as the isotopes in cores taken through the ice caps in Greenland and Antarctica bear on hundreds of thousands of years. Various devices have been devised for weeding out the magnetic signal. The Earth's magnetic field has little effect on meteorites before they land so that their cosmogenic signature is a more direct measure of solar activity than are the isotopes in ice, sediments or wood. The titanium 44 (^{44}Ti) activity of 15 meteorites which fell during the last 2 centuries, for example, provides this kind of information because it is produced by the action of GCRs.

Besides their value in tracing long-term changes the cosmogenic isotopes (that is isotopes produced by GCRs) are clues to the Sun's present vigour. In November 2003, for example, the beryllium 7 (^{7}Be) content of the air at Thessaloniki in Greece fell markedly 11 days after a gust of solar wind was emitted by the Sun.[25] The isotope ^{7}Be has a half-life (t½) of 53 days, that is to say the original number of ^{7}Be atoms is halved by radioactive decay in 53 days, whereas ^{10}Be has one of 1,500,000 years. This makes ^{7}Be a sensitive measure of very brief events.

In other words, the evidence stored in the Antarctic ice records events on the Sun hundreds of thousands of years ago, the air in Thessaloniki tells us about something that happened on the Sun a few days ago. Together they partly plug the gap between what we will learn from the GENESIS space mission about events at the Sun's creation 4.5 billion years ago, and the images in our telescopes of what was happening on the Sun eight minutes ago. A history of the Sun is falling into place.

Chapter 2
Inside the Sun

In 467 BC a burnt-looking stone the size of a *wagon load* fell from the sky near the river Aegos, in Thrace. Wagon load is a nice touch, for it vividly conveys that the object was large and its fall spectacular. The astronomer Anaxagoras very reasonably concluded that it had fallen from the Sun and by implication that the Sun was a glowing mass of rock or, according to one account, a red-hot lump of metal.[1]

Rock or metal, Anaxagoras was imprisoned for impiety, as he had contradicted the current doctrine that the Sun, like the Moon, was a deity, and only the intervention of his pupil, the great general and politician Pericles, ensured the death sentence was commuted to a fine of five talents and exile.

Anaxagoras was on to many other promising items of solar astronomy: in the words of Hippolytus[2] he claimed that *the Sun and Moon and all the stars are fiery stones* (although he understood perfectly well that the Moon shone by reflected light*)*, that *the Sun is larger than the Peloponnese,* that *the Sun is eclipsed when the new Moon goes in front of it*, and that *the Milky Way is a reflection of the light of those stars that do not get their light from the Sun.*

That he was a flat-earther and held several cranky ideas in biology as well as cosmology should not detract from his realisation that the solar system was all of a piece and that it could be explained on the basis of everyday experience. And in a rather roundabout way he was right to link meteorites to the Sun. For by far the commonest type of meteorite falls are the carbonaceous chondrites, which are fragments of asteroids that formed early in the history of the solar system and have not been melted since, so that they have retained the character of the solar nebula. And they are chemically very similar to the Sun's photosphere, which, though consisting almost entirely of hydrogen and helium, also contains traces of oxygen, nitrogen, neon, carbon, iron, silicon, magnesium, sulphur, aluminium, sodium and calcium.

Anaxagoras' hot stone model of the Sun did not flourish during the next two millennia, as there were many who preferred something more ethereal. Even the notion of falling meteorites fell into disrepute. The world began to be persuaded in 1492 when a meteorite weighing 127 kg (a chondrite) fell near Enisheim, in Alsace; but, although it was taken as a favourable sign from God (the Emperor Maximilian duly won an impending battle), the view still prevailed that stones do not drop out of the sky. More convincing was a fall in Siena in 1794, which was witnessed by

C. Vita-Finzi, *The Sun: A User's Manual,*
doi: 10.1007/978-1-4020-6881-2_2, © Springer Science+Business Media B.V. 2008

many townspeople, but in neither case was a connection drawn between the meteorite and the Sun. But it was still possible for Thomas Jefferson, the third US president (1801–1809) and often described as a man of the Enlightenment, to remark that it was easier to believe two Yankee professors had agreed than that stones fell from the sky.

Despite the lack of any direct evidence, the notion of a hot Sun in the process of cooling regularly surfaced, helped no doubt by the continuing debate about the mechanism responsible for heating the Earth's interior. Miners and geologists had long known that temperatures increased with depth. Had the Earth started out hot and gradually cooled? Applied to the Sun this line of argument gave Isaac Newton an age for the Sun of 50,000–75,000 years, which was consistent with some interpretations of the Biblical narrative but still left open the question of how the Sun became hot in the first place.

Little progress was made in understanding what powered the Sun until the middle of the 19th century, when it was proposed by William Thomson (later Lord Kelvin) that the Sun gained heat from the kinetic energy released by the infall of meteors or entire planets. Kinetic energy is the energy derived from motion, which converts to heat when the object hits the buffers. It was soon clear that there were too few meteors or planets for the purpose and that there was no evidence that the process had slowed down once the supply of meteors and planets had been exhausted.

Kelvin then adopted the suggestion of H. von Helmholtz that the energy was derived from the progressive contraction of the Sun. (This hypothetical process is due to be reversed when, as discussed later in this chapter, the Sun goes through a Red Giant period and swells monstrously.) By postulating that the Sun originally had a diameter 220 times larger than now and had shrunk at about 20 m a year, he came up with a maximum possible age for the Sun of 100 million years.

In the fifth edition of his authoritative book *The Earth* published in 1970, the geophysicist Harold Jeffreys showed that gravitational contraction[3] was 10,000 times more effective as a source of heat for the Sun than any known chemical reaction. Jeffreys concluded that contraction could have yielded the present rate of radiation from the Sun for 25 million years, a quarter of Kelvin's estimate; the corresponding value obtained for chemical processes was a meagre 3 million years. But he also noted that the period of pulsation of the variable star δ Cephei was changing at a rate 1/170 of what was to be expected if the star were indeed driven by contraction. Variable stars are discussed later; what matters here is that (according to Jeffreys) the frequency at which they vary in brightness is related to their density (strictly $1/\sqrt{\text{density}}$), so that it should change as the star contracts and becomes denser. The change predicted for δ Cephei was 17 s a year whereas the observed change was 1/10 s. Therefore the contraction argument was invalid. Other Cepheid variables gave a similar result. (Cepheid variables, like many of us, brighten rapidly and then gradually grow dim.) Ergo, the contraction model for the Sun was suspect.

Jeffreys always insisted that science should have a sound mathematical basis, and is often remembered for his opposition to the concept of continental drift on the grounds that the quantitative evidence showed it to be impossible and that it was an example of what he termed 'the reckless claims that get into the newspapers'. (Some years later the successful successor of continental drift, plate tectonics, was reputedly dismissed by a reviewer as the sort of thing people say at cocktail parties.) But Jeffreys was not invariably a reactionary: his calculations showed that a new explanation was required to explain the Sun's energy output. Astronomy, like geology, was starting to show that the short timescale of Kelvin and von Helmholtz was in error.

A new source of solar energy

Natural radioactivity had been discovered in 1896, and by 1903 it was established that radium salts generate heat without cooling down. As far as the Earth was concerned the fission of uranium, thorium and potassium in rocks was soon seen to provide a plausible source of heat that would greatly extend the geological timescale. Ernest Rutherford, pioneering nuclear physicist, tells of a lecture in 1904 which was attended by Lord Kelvin himself at which Rutherford raised the possibility that the newfangled source of heat made possible a greatly extended age estimate for the Earth. Lord Kelvin, he said,

> had limited the age of the earth, provided no new source (of energy) was discovered. That prophetic utterance refers to what we are now considering tonight, radium! Behold! the old boy beamed upon me.

By 1917 radioactive dating had shown that the Earth was about 2 billion years old. The age of the Sun was unlikely to be 20 million years. Yet heating the Sun by radioactivity presented problems. If the Sun were composed entirely of pure uranium it would radiate at roughly the present rate but, whereas the Earth is rich in radioactive elements, the Sun is not. By 1925 the young Cecilia Payne had shown that the solar atmosphere was rich in hydrogen. The celebrated astronomer Arthur Eddington riposted that this might be true of the Sun's exterior but not its interior; yet by 1932 it was widely accepted that hydrogen made up 1/3 of the Sun.[4] As the tubeworms of Chapter 5 will demonstrate, this is not the last time a young female researcher is snubbed by a scientific grandee only ultimately to triumph.

A persuasive explanation for the Sun's luminosity came from the combined teachings of relativity and nuclear physics. The Sun had already served to link Eddington to Einstein. In May 1919 Eddington led an expedition to the island of Principe, off west Africa, to test one of the corollaries of general relativity, which Einstein had published in 1915: that light would be deflected by gravity. A parallel expedition went to Sabral in NE Brazil. The expeditions were timed to coincide with a total solar eclipse so that they could observe the effect of the Sun on light from stars beyond it. The path of the eclipse would lie near the Equator, hence the choice of the two sites on the Atlantic. Measurements on 12 stars from both locations were compared with reference photographs taken at Greenwich in the UK in January.[5]

Einstein had predicted a deflection of 1.745 seconds of arc. Eddington's measurements gave a generalized measurement of 1.64. Critics have suggested that Eddington discarded 2/3 of the data to ensure a favourable result, but observations during the 1922 eclipse in Australia, and modern studies employing distant radio sources such as quasars, rather than stars, have supplied ample confirmation of the predicted effect. Moreover an exquisite test of relativity comes daily from what is now the familiar technology of satellite positioning. As shown in Chapter 7, the GPS satellites on which it depends are at an elevation where the effects of both general relativity and special relativity have to be allowed for in order to provide the system with the nano-second (ns, $1/10^9$ s) accuracy it requires if errors are not to accumulate at a rate of 10 km *a day*.[6] But, fairly or not, this extraordinary achievement can in no way compete with the results of Eddington's expedition for its impact on the public.

It was now the turn of relativity to repay the compliment and resolve a solar riddle. Einstein had shown in 1905 that mass and energy were interchangeable, and the great heat and pressure within the Sun was recognized as propitious for the release of energy from the conversion of hydrogen to helium. In 1920, Eddington had argued as follows: the mass of a hydrogen atom is 1.008 which multiplied by 4 (as four H atoms might combine to make an atom of helium) is 4.032, well in excess of the 4.004 mass of an atom of the commonest form of helium, He-4. The balance, following Einstein, would be available as energy. If a mere 5% of the Sun consisted of hydrogen its transformation to helium would liberate ample energy. But at the time the Sun was thought to have a composition similar to the Earth's, and the interaction of subatomic particles was little understood.

By 1928, however, analysis of the Fraunhofer lines in the solar spectrum (see Fig. 1.7) had shown that hydrogen is the most abundant element in the Sun's atmosphere. The Sun, in short, was not just a hot version of the Earth. Of the plausible pathways the first to be proposed is known as the carbon-nitrogen-oxygen (CNO) cycle. According to a story spread by George Gamow, who had organized a conference in Washington, DC, on energy generation inside stars, the physicist Hans Bethe worked out the CNO process on the train journey home to Cornell, where he was based, and, what is more, he did so, as he had intended, before the steward called the passengers to dinner. (Bethe enters into another Gamow story as the third contributor to a early paper on the Big Bang by Gamow and his student Raph Alpher, having been invited to co-author it purely because Gamow wished to complete the authorial lineup with the first three letters of the Greek alphabet.)[7]

The CNO cycle is now thought to operate effectively only in stars with a greater mass than our Sun and a correspondingly hotter interior, and contributes only 1.5% of the Sun's energy output.[8] The scheme now accepted for the Sun, also worked out by Bethe, is the proton-proton (*p-p*) chain. The process is in three stages culminating in the fusion of two nuclei of helium-3 to form a nucleus of helium 4 and the release of energy. The cycle uses up about 700 million tons of hydrogen a second, but the Sun is sufficiently massive for this process, which began about 4,500 Myr ago, to continue for a further 4,500 Myr.

Fusion in the core for both the CNO and the *p-p* sequences was possible only when temperatures had been rendered high enough by gravitational energy, perhaps 30 Myr after the Sun's birth.[9] Kelvin and other 19th-century physicists were therefore partly right in appealing to gravity as the driving mechanism.[10]

The solar onion

The standard model requires the following arrangement: an inner core where the fusion takes place, an intermediate zone where heat radiates outwards, and an outer zone where the heat is conveyed by convection (Fig. 2.1). As we have already seen, the convecting zone underlies the visible solar surface or photosphere which in turn is overlain by the chromosphere and the corona.

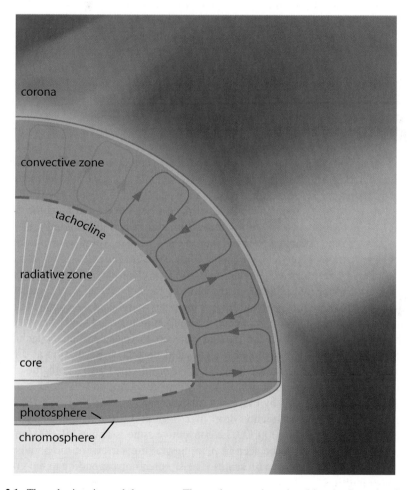

Fig. 2.1 The solar interior and the corona. The nuclear reactions that drive the Sun take place in the core. The resulting energy is transmitted by radiation to the tachocline; despite travelling at the speed of light an individual photon may take a million years to make the journey. Temperature drops from about 7 million degrees Celsius near the core to 2 million degrees Celsius at the tachocline, where the Sun's magnetic field is probably generated, and 5,700°C at the visible surface of the Sun (the photosphere). Temperatures in the corona rise to over 1 million degrees Celsius

Since its launch in 1962 helioseismology provides a complementary angle of attack on the problem of solar energy. For instance, its data have already refined estimates of the depth of the convective zone to the very specific value of 0.713 (i.e. a little over 7/10) of the solar radius. The content of helium in the convective zone comes out at 0.248 (i.e. almost 25%) compared with estimates from astrophysics of about 0.275; and, most important for the neutrino problem, the seismic data are fully consistent with the standard model for the solar interior.[11] The probable explanation for the shortfall in neutrino capture goes back to a proposal made in 1969: that the various kind of neutrino produced in the Sun change character ('flavour') as they travel to Earth, and only some of them can be detected with existing methods. In short the problem lay not with the SSM but with what has been called[12] the multiple personality disorder of neutrinos.

The solar core is the reactor where the fusion processes take place. It occupies ¼ of the distance from the Sun's centre to its visible surface or photosphere. The density of the core is about 14 times that of lead; temperatures are in the order of 15 million degrees Celsius and pressure equivalent to 300,000 million of our atmospheres. The energy created within the core takes the form of gamma rays as well as neutrinos. The gamma rays interact with atoms on their way out to space and are converted into a much larger number of photons.

Only 1/50,000,000,000 of the thermal radiation generated in the core manages to escape and it may take a million years to do so.[13] The first stage of the journey, as we saw, is by radiation. The gas here is highly ionised, that is to say electrically charged because its constituent atoms have been stripped of electrons, the outcome being a plasma. (Sometimes called the fourth state of matter after solids, liquids and gases, plasma is by far the commonest of the four in the Universe, yet unknown to most of us except perhaps, and aptly enough, in the context of some attempts to achieve fusion in the laboratory in order to generate pollution-free electricity.) Temperatures are a modest 7 million degrees Celsius at the base of the conducting zone. Radiation is inefficient in the sense that heat builds up at the outer margins of the conducting zone, resulting in steep temperature gradients, whereupon heat loss is now by convection.

The boundary between conducting and convective zones is called the tachocline. The Sun's magnetic field is probably generated here because plasma at different depths and latitudes moves at different rates. As within the Earth's core, the circulation of conducting material drives some kind of dynamo, and its field is aligned broadly along the N-S axis by the Sun's rotation.

Temperatures at the tachocline drop to 2 million °C, and at the surface of the convective zone to about 6,000°C. The convective zone is about 150,000 km deep. The convective pattern can be seen in plan view at the surface of the photosphere. The process operates at two main scales. The smaller, corresponding to the granulation illustrated in Fig. 2.2, consists of units about 1,000 km across. The larger is represented by supergranulation cells perhaps 30,000 km across and with surface movements of about 400 m/s. It is thought that cells deeper within the convective zone are at even larger scales. Spectroscopy confirms that the granule centres represent rising material because their spectra are blue-shifted, that is to say moving towards the observer, whereas their margins are shifted towards the red, and therefore moving away, at rates of about 1 km/s.

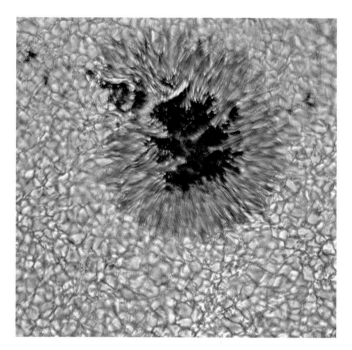

Fig. 2.2 Sunspot and granules on the photosphere (Photo credit: Royal Swedish Academy of Sciences) The granules are the tops of convection cells 1,000–2,000 km in diameter in the convective zone

The photosphere has a density about 1/100 that of our atmosphere but it is very opaque because some of its hydrogen atoms are very efficient at absorbing light.[14] The opacity contributes to the build up in temperature as it impedes the passage of photons, the elementary particles that carry electromagnetic radiation of all wavelengths, including radio waves, UV light, visible light and gamma rays.

The surface of the photosphere is the source of most of the radiation emitted by the Sun and is the visible solar disk. Its texture is best known for its sunspots. Their study by Galileo and his contemporaries was of profound significance not only for what they revealed about the Sun itself and for what they did to Aristotle's reputation but also because they showed that a rotating Earth – an integral part of the Sun-centred model of the solar system advanced by Copernicus in 1542 – was a wholly reasonable proposition. (Copernicus' great work remained on the Catholic Church's list of forbidden books until 1835. Galileo fared even worse: the Vatican did not rehabilitate him until 1992, that is to say 359 years after he was accused of heresy, probably the slowest bit of peer review in history.)

It was in the photosphere that helium was first identified during the 1868 eclipse as an emission line for an unknown element. The line could not be produced in the laboratory. Norman Lockyer, who observed the line in the same year, named the new element helium after the Greek name Helios for the Sun god. The gas was not discovered on Earth until 1895. It makes up only 1/2,000 of the Earth's atmosphere and is produced on Earth by radioactive decay, but it has turned out to be the second most abundant element in the universe after hydrogen thanks to its primeval abundance created in the Big Bang and its production in the stars by the fusion mechanism that resolved the riddle of the Sun.

The chromosphere, 5,000 km thick, is named after the reddish arcs which are sometimes observed around the Sun during eclipses (Fig. 2.3) and which are produced by emission from the red hydrogen H alpha (Hα) emission line. Hence it can be observed at times other than eclipses by viewing (or photographing) it through an Hα filter, which eliminates the light from the chromosphere. The spicules at the surface of the chromosphere are thought to lie above the boundary between supergranules in the photosphere and thus perhaps represent escaping jets of gas. Why the chromosphere should be 4,000° hotter than the photosphere remains a puzzle. One suggestion is that it is heated by turbulence and the action of shock waves.

What may be considered the Sun's atmosphere, the corona, is composed of tenuous gases with temperatures which average 1 million degrees Celsius but can attain 2 million. It is far less bright than the photosphere, which is why in pre-instrumental times it was visible only during eclipses when the Sun was blotted out. We know its temperature because some of the emission lines in its spectrum are those of atoms which have been strongly ionized under the action of high temperatures and consequently carry an electric charge. X-ray and extreme UV imaging reveal dark holes which, like sunspots, have lower temperatures than their surroundings.

UCAR/NCAR/High Altitude Observatory   Solar Corona – 11 July 1991

Fig. 2.3 Prominences from chromosphere and streamers from corona revealed during solar eclipse (Courtesy of NASA and UCAR/NCAR) showing prominences in red (H alpha)

The corona emits energy mainly in UV and X-ray wavelengths. It also experiences coronal mass ejections (CMEs: Fig. 2.4), which can attain speeds of 200–2,500 km/s at 1 AU, and, as will be shown in Chapter 7, have serious practical consequences for us on Earth. Its everyday association with the solar wind appears otherwise uneventful, but as often happens, averages can be deceptive. Thus, although the solar wind blows at about 400 km/s from the solar equator its speed from coronal holes near the poles is nearer 800 km/s.

The region of space dominated by the solar wind is the heliosphere, whereas that dominated by the Earth's magnetic field is known as the magnetosphere (Fig. 2.5).

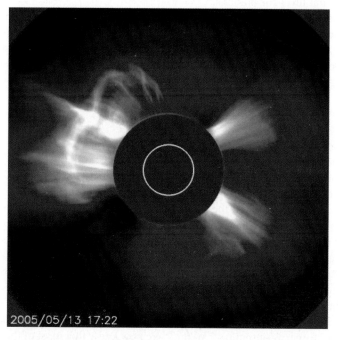

Fig. 2.4 The coronal mass ejection (CME) on 13 May 2005 was heralded by large flare. The CME hit the Earth's atmosphere after 33 h. Protons had started arriving a few hours after the flare; when they peaked, cosmic flux fell, the Earth's magnetic field was distorted, and auroras were seen at lower latitudes than usual (Courtesy of NASA). In the image the Sun is masked to reveal the CME

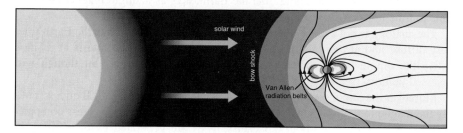

Fig. 2.5 The collision between the Earth's magnetosphere (in blue) and the heliosphere, the volume filled with the solar wind, is marked by a bow shock

The heliosphere extends about 100 AU from the Sun and thus well beyond the orbits of the planets in our solar system. Its outer limit is the heliopause. In 2004 the Voyager 1 spacecraft, which together with Voyager 2 was launched in 1977, crossed the termination shock, where the supersonic solar wind drops to subsonic speeds in response to pressure from the interstellar wind. It was then 94 AU from the Sun. and its messages took 14 h to reach the Earth at the speed of light. Traveling at about 17 km/s, Voyager 1 will cross the heliopause in a further 10–20 years. The spacecraft incorporates substantial shielding of sensitive instruments against radiation: near the Jupiter swing-by the dosage was equivalent to 1,000 times the lethal level for humans.

The aurora borealis (northern lights) and the aurora australis (southern lights) encircle the poles and are displayed when energetic electrons in the solar wind drawn in by the magnetic poles excite oxygen and nitrogen in the atmosphere 100–400 km above the ground (Fig. 2.6). In other words the solar embrace goes beyond heat and light and envelops the Earth in a cloud of charged particles.

The magnetic Sun

In 1833 John Herschel (son of William) proposed that sunspots were associated with magnetic fields. This suggestion was not given substance until 1908, when George Ellery Hale exploited the Zeeman effect and detected strong vertical magnetic fields at the spots. These fields may account for the relative coolness of the spots – we are talking about 4,000°C compared with about 6,000°C – because they would act as barriers to the horizontal movement of gases and thus inhibit the convection that brings up heat from below the surface.[15] But the assumption that sunspots are a measure of solar activity and therefore of its luminosity still holds because their cooling effect, amounting to 1 W/m², is outweighed by that of the larger, brighter zones (plages or faculae) with which sunspots are associated and which increase energy output by 2 W/m².[16] We now know that most of the temporary features on the surface of the photosphere, and not just sunspots, are in some way shaped by magnetic fields and in their turn generate other magnetic fields even though only 1/10,000 of the energy flowing out from the core is used for these tasks.[17] And the Sun as a whole, like the Earth, acts as a single magnet and from time to time reverses its north and south poles.

The analogy is not a close one. The Sun's magnetic poles, unlike the Earth's, are very diffuse and at their sharpest at solar minimum; when the Sun is at its most active the dipole breaks down and then reasserts itself bit by bit though with reversed polarity (see Chapter 3).

Moreover the Sun's general field is complicated by numerous local effects, notably the magnetic fields at sunspots, which typically measure 2500 Gauss compared with the Earth's field of about 0.3 Gauss. They are generated by the differential rotation of the Sun with respect to latitude: one revolution takes 25 days at the equator and 28 days at mid-latitudes. According to H.W. Babcock this disrupts the

Fig. 2.6 Aurora borealis over Lac du Flambeau, Wisconsin (Courtesy of NASA – in global-warming.accuweather.com/2007/03/01). The strong red and green emissions in auroras arise from emissions from oxygen atoms from oxygen molecules which have been ionised by UV radiation

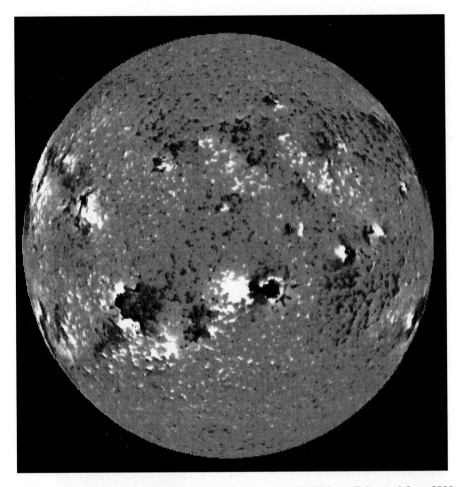

Fig. 2.7 Magnetogram revealed by imaging the Sun at Ca II 854.2 nm light on 6 June 2000 (Courtesy of National Solar Observatory, Kitt Peak)

quiescent dipole field and eventually creates "knots" of strong localised fields some of which penetrate the photosphere to form sunspots.

This magnetic complexity is best visualized by using magnetograms (Fig. 2.7), which are constructed by scanning across the solar disk and measuring the intensity variation of a particular spectral absorption line profile as revealed by Zeeman splitting. The convention is to show neutral areas gray, strength of field pointing away the observer in shades of black and strength of field pointing towards the observer in shades of white. North is at the top but west is to the right, the convention in astronomical images. As these are magnetic (and not thermal) images, sunspots may appear black or white according to their magnetic polarity, with positive (white) corresponding to north and negative (black) to south. Many major sunspots come in pairs.

Other suns

On 17 February 1600 Giordano Bruno was burned at the stake in what is now the Campo de' Fiori or Field of Flowers in Rome. A statue commemorates the barbarity. It is said that religious politicians objected to the statue (by Ettore Ferrari) when it was first proposed by Victor Hugo, Herbert Spencer, Henrik Ibsen and others in 1885. It was finally erected in 1889.

Bruno was a Dominican who attracted the attention of the Inquisition by his unorthodox religious and philosophical views. He renounced his vows and fled from Italy in 1576. Bruno had moved from one country to another to avoid trouble, first, to France; then to Germany; and finally to England, where in the course of a 2-year stay he published a number of books. They included O*n the Infinite Universe and Worlds* (1584), which maintained that the stars were like our Sun and that they were circled by planets inhabited by intelligent beings.

The Church's objections to Bruno's ideas were of course rooted in the Scriptures or more precisely in the lack of any scriptural mention of other worlds and their inhabitants. Even Galileo found the notion of extraterrestrial life ludicrous, although you can never be sure when he is teasing or merely avoiding unnecessary disputes. Discussing the suggestion that the Moon was inhabited, for instance, Galileo raised the problem of explaining how its inhabitants could be descended from Adam. Did they too bear original sin? This religiosity was endorsed by more practical matters. Galileo saw that there were no clouds on the Moon, and concluded that therefore there was no water. He also thought it likely that any former life would have been wiped out by the present pattern of scorching heat alternating with freezing temperatures that results from the absence of any lunar atmosphere.

Bruno returned to Italy as tutor to a Venetian called Mocenigo but before long he was handed over to the Inquisiton in Rome and, after prolonged imprisonment and torture, convicted of hcresy. The grounds for his execution are not clear, as his writings included support for the Copernican model of the Universe and blasphemous views on the virginity of the Virgin as well as his ideas on other worlds. There are now revisionist accounts of his life, as there are of Galileo's, which deride the popular view of Bruno as a symbol of the intolerance of authority in the face of new ideas, dismiss his significance in the history of astronomy, and overlook the little matter of the torture and burning: 'almost all of his misfortunes were brought down upon himself without the Inquisition's help … his astronomical writings reveal a poor grasp of the subject on several important points … he vocally espoused (but apparently did not really understand) Copernicanism … his *On the Infinite Universe and Worlds* appeals to many today because of its apparent resonance with the deeply held conviction that life exists elsewhere in the Universe … a kind of culture hero instead of a footnote in books on Renaissance philosophy'.[18]

By the 1640s the idea of other planetary worlds had surfaced again. Descartes concluded that the Sun and the stars were similar because they were all luminous; his followers completed the analogy by adding inhabited planets to the panorama.[19] And, whereas, Galileo and Johannes Kepler rejected any Brunoan notion of a plurality of solar systems, though more from caution than conviction, Henry More wrote (in 1646) of fixed stars around which spun Earths, Jupiters and Saturns.[20] By the close of the 18th century William Herschel could maintain that the Sun, like other stars, underwent changes in brightness and thus modified the conditions prevailing on its planetary retinue, and he concluded that the cool dark surface glimpsed through sunspots might well be inhabited. If stars are suns and suns are (like our Sun) inhabitable, he mused, 'what an extensive field for animation opens itself'.[21]

The most durable attempt to place the Sun in the context of other stars was by the Dane Einar Hertzsprung and the American Henry Norris Russell, who, working independently, came up with very similar results. A version of the resulting Hertzsprung-Russell (or H-R) diagram is shown in Fig. 2.8. It is a summary of a plot of the 10,000 brightest stars seen from Earth according to their temperature and intrinsic brightness or luminosity. The Sun lies in the main sequence and it is classed as a G2 star.

Note that, contrary to current usage, but respecting the original diagram, temperature decreases away from the origin of the graph i.e. from left to right. Brightness has been substituted for magnitude in the original and temperature for spectral class. Magnitude is a concept that derives from the work of Hipparchus, who divided his catalogue of about 850 stars into six classes according to their brightness. (Patrick Moore has noted, as in golf handicaps, the lower the number the greater the apparent magnitude of a star). Hipparchus' six category scale has been extended in both directions and refined using decimals, and the apparent magnitude he had to work with can be converted to absolute magnitude by allowing for the star's distance from Earth, which of course is information not available to him. The accepted conversion is how bright the star would appear if it were 10 parsecs (that is about 32 light years: a parsec represents a parallax against the background of fixed stars of 1 s of arc) away from the Earth. The Sun's apparent magnitude (m) of -26.72 then becomes an absolute magnitude (M) of 4.8; that of the Dog Star (Sirius α) changes from -1.44m to 1.47M.

Temperature can be derived from the star's colour or spectral class by imaging it through different filters, usually UV, blue and yellow. Red stars are the coolest, white ones the hottest. The accepted categories, from hottest to coolest, are O-B-A-F-G-K-M (Table 2.1). Three main groupings emerge: the cold giants, the hot dwarfs and the constituents of the main sequence ranging from bright blue stars to faint red stars. Some preliminary conclusions were soon drawn from these basic observations. For instance, the luminous red stars had to be very large because they were bright despite being cool. The paucity of red giants (as they are now called) suggested that as they are in short supply they must be in a stage of rapid evolution.

In 1925 Cecilia Payne compared the spectrum of the Sun to that of other stars to show that virtually all bright, middle-aged stars have the same composition.

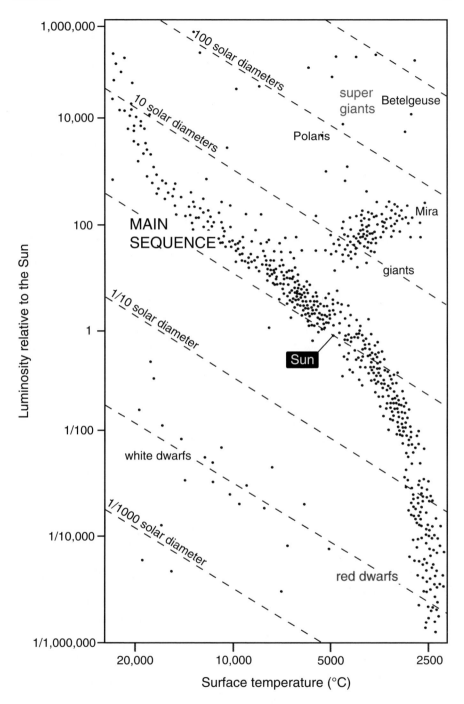

Fig. 2.8 A Hertzsprung-Russell diagram, which is a summary plot of the brightest stars seen from Earth according to their temperature and intrinsic brightness or luminosity. Note that the diameter scale (dashed lines) is logarithmic so that each step represents a tenfold increase. The Sun is near the middle of the main sequence

Table 2.1 Star colour groups and temperatures

Code	Colour	Approx. temp. (°C)	Example
O	Blue	28,000–60,000	Zeta Orionis
B	Blue white	10,000–28,000	Rigel
A	White	7,500–10,000	Siorius A
F	Yellowish white	6,500–7,500	Canopus
G	Yellow	5,000–6,000	Our Sun
K	Orange	3,500–5,000	Aldebaran
M	Red	2,000–3,500	Betelgeuse

Middle-aged is a notion derived from the H-R diagram. As has often been observed, the HR diagram is like a family picture which includes new babies, teenagers, parents, and so on.

This cosy image conveys the key point of the H-R diagram, namely to indicate the likely fate of stars on the graph. It is commonplace in science (and sometimes mistaken) to assume that related features of different age which exist side by side form part of a time sequence: thus babies, teenagers and adults, or gullies, deep valleys and wide alluvial floodplains. So with stars at different locations in the main sequence, and thus the Sun, which is how predictions of its lifespan and development come to be made.

The predictions state, simply, that the Sun will continue to rise in temperature for a further 4,500 Myr, when 10% of the hydrogen in its core will have been converted to helium and the core will collapse under gravity.[22] The resulting heat will swell the Sun so that its surface is far enough from the heat source at the centre to cool from white hot to red hot. It will remain a red giant, its diameter roughly the same as the present orbit of Mars, for several hundred million years; then, after several cycles of expansion and contraction and when its fuel is exhausted, it will shed its outer layers and collapse, to become an extremely dense white dwarf, with the diameter of a typical planet, and once all its energy has been radiated away, in due course a carbon-rich, cold, dead, black dwarf.

The search for other suns goes beyond the issue of age to consider behaviour both for their bearing on the star in question and for any insights it might provide into the Sun's sunspots and other semi-periodic and non-periodic surface disturbances. In 1638 Mira, a star in Cetus (o Ceti), was firmly shown to be a variable star, that is one which varies in light output. Variability is either intrinsic, when it is due to changes in the luminosity of the star itself, or extrinsic, when it is due to its rotation or to eclipsing by another star (eclipsing binary).

Intrinsic stars fall into two groups: pulsating, which as the name implies expand and contract periodically, and eruptive or cataclysmic variables, which have occasional violent outbursts. The four main types of pulsating variable stars are Long-period (80–1,000 days), RV Tauri (30–100 days) Cepheids (1–70 days), and RR Lyrae (0.2–1.0 days). A fifth group, the semiregular variables (30–1,000 days), are giants and supergiants which show periodic behaviour alternating with irregular spells. The eruptive variables include supernovae, for which no periodicity is reported, novae (1–300 or more days), binary systems with a Sun-like star and a white dwarf, recurrent novae (1–200 or more days) with two or more minor out-bursts, and dwarf novae, binary systems with a Sun-like star and a white dwarf sur-rounded by an accretion disk. Rotating stars show changes in light output which could be due to light or bright spots or patches. Eclipsing binaries have an orbital plane close to the line of sight of the observer.

In addition variable output is displayed by flare (or UV Ceti) stars, which are red main-sequence stars displaying shortlived, localized outbursts, and irregular varia-bles, which include most red giants.

By 1781 eight variable stars were known of which four (like Mira) were periodic. By 2004 38,622 variable stars were known, but attempts to secure total numbers of periodic stars are hampered by the likelihood that those with periods longer than that of the survey may be missed. Nevertheless two groups have been provisionally identified: those with periods shorter than 50 days (mostly less than 10 days) and those with periods of 100–450 days. Mira is periodic, with a period of about 332 days; it is perhaps less luminous than our sun near the solar minima discussed in the next chapter but up to 1,500 times as bright near the maximum of its cycle.

Other solar systems

Some of the extrinsic category owe their variability, as we saw, to their being repeatedly eclipsed by another star. The question now arises whether, to quote Oscar Wilde, the body that interposes itself is that of a planet rather than another star. Bruno had hazarded the guess that other suns would be so endowed. That might make a convenient criterion for planetary status.

Astronomers have long speculated that planets would form from the slowly rotating disc-shaped cloud of dust and gas that is believed typically to surround a newly formed star. A variety of planets might emerge, such as gas giants, icy bod-ies or indeed rocky planets like the Earth. A gas giant such as Jupiter shelters Earth-like planets from cometary impacts with its gravity, and it has been sug-gested that the search for a civilization like ours could begin with a trawl for Jupiter-sized bodies.[23]

The first extrasolar planet (or exoplanet) to be observed, using Doppler tech-niques, was orbiting the star 51-Pegasi and had a mass similar to that of Jupiter.[24] The orbit had a very small radius, 1/20 AU, but even a gas giant could survive during the lifetime of a main sequence star if it had started out further away and gradually migrated to this orbit.[25] By May 2007 the catalogue had grown to 236 exoplanets around nearby stars, detected by Doppler and also by transit methods where the pas-sage of the planet in front of the star leads to a temporary fall in the star's brightness.

Explicitly or not, the study of other worlds is motivated in part by the hunt for alien life. A planet around the main sequence star 70 Virginis, for example, has a mass about nine times that of Jupiter but an average surface temperature of 85°C. Besides temperature the crucial test of habitability is often deemed to be the presence of water, although in view of the growing evidence for life in extravagantly extreme environments it seems old fashioned to insist on this criterion. At all events there was much excitement when, in 2007, water was detected on the planet HD 189733b, 63 light years away in Vulpecula and with a diameter 1.25 that of Jupiter. The planet was discovered by the transit method as it dims the star's light by 3% every 2 days when passing in front of it. The planet orbits at a distance equivalent to 3/100 of an AU; its atmospheric temperature hovers around 700°C, which is why its is known as a 'hot gas-giant' or 'hot Jupiter'. Analysis of its atmosphere using the Spitzer Space Telescope demonstrated the presence of water vapour on the planet.[26] The same year saw the discovery of a third planet orbiting Gliese 581. Known as Gliese 581 C, it is of great interest to exobiologists because it occupies the habitable zone around its parent star, with a mean temperature between 0°C and 40°C, and its mass suggests it could be rocky.

By November 2007 there were 189 known planetary systems (including our own) containing at least 222 known planets.[27] Following Bruno we could usefully apply the term sun to their parent stars, just as the uncapitalised term moon serves well in cataloguing satellites, such as the moons of Jupiter, within our solar system.

Chapter 3
The Changeable Sun

In *Sunshine*, a film set 50 years into the future, written by Alex Garland and released in AD 2007, a team of astronauts is sent to re-ignite the dying sun. The plot is a little at odds with the standard model, which, as we saw, predicts that the Sun will soldier on for atleast another 4 Gyr. But it makes by implication two telling points omitted from many astrophysics textbooks: that our knowledge of the past 4.6 Gyr of solar history is very incomplete and mostly circumstantial, and that (as we saw) the conventional view of the Sun's future is grounded in a model which is still being refined.

Although portions of Solar System history can be pieced together from meteorites, interplanetary dust particles (IDPs) and the battered rock record of the inner planets Mercury, Venus, Earth and Mars, the history of the Sun itself is confined to the four centuries of telescopic observation or, at best, the two millennia during which Chinese, Korean and European astronomers have kept a tally of its sunspots.

There is some information on solar flares and cosmic rays over the last 1 Gyr to be recovered from lunar surface deposits, and on solar wind activity over the last 8,000 years to be inferred from tree rings and 300,000 years from cored polar ice. Direct observation by current or recent space missions documents the last 40 years. That leaves over 3 Gyr lacking the information that is required for testing competing models for the Sun's past behaviour.

The young Sun

According to prevailing theory, soon after its birth and on the assumption that its hydrogen store was being depleted at the estimated rate of 700 million tons per second, the energy emitted by the Sun was perhaps 70% of its present value. By a principle known as Stefan's Law the surface temperature of a planet varies by ¼ power of the solar energy falling on it, so that the Earth's mean temperature would then have been 91% of its present value, and as the law is expressed on the Kelvin scale this means $-15°C$ compared to the present value of $+10°C$. Mars too must have felt the impact of the low solar output. As its orbit is 1.52 AU, and as energy receipts fall off as the square of the distance, the mean temperature on Mars would have been $-77°C$, compared with $-55°C$ today.[1]

C. Vita-Finzi, *The Sun: A User's Manual*,
doi: 10.1007/978-1-4020-6881-2_3, © Springer Science+Business Media B.V. 2008

There follow many geological and biological puzzles, discussed in the next chapter, which can be circumvented at a stroke by arguing that the Sun is only 6,000-years old as recorded by Biblical chronology – or at any rate the Victorian interpretation of it. The debate is enlivened by the use of the 'young Sun' to mean both the creationist version and our middle-aged Sun as an infant.

Even if we do not shut our eyes to the superabundant physical evidence around (and beneath) us for billions of years of solar history there is scope for an alternative view of the early stages, namely that (contrary to the standard model for main-system stars) the Sun did not become brighter but rather lost brightness by vigorously shedding material as solar wind. The suggestion has been tested by observations of a nearby Sun-like star, π 01 Ursa Majoris, estimated to be only 300 Myr old, and it was found to lose mass at a rate far too low to satisfy the hypothesis.[2] So we are stuck with a wan youngster.

If the Sun continues to evolve as a main-sequence star the progress of nuclear fusion will continue to increase the ratio of helium to hydrogen in the core. Helium is denser than hydrogen; higher core density means more efficient nuclear fusion and therefore higher core temperature, and thus a brighter Sun until nuclear fusion has converted all the hydrogen in the Sun's core into helium.[3]

We are of course having to make the usual assumption that all main-sequence stars have a similar evolution. By this token, the Sun's future can be reviewed by looking at older stars of similar lineage. Some of the red-giant stars in the globular cluster NGC 104 (47 Tuc), similar in mass to our Sun and about 15,000 light years away, have used up their hydrogen and are in the final stages of the red-giant phase, losing about one millionth of their mass each year. Others, at an earlier stage in the red-giant phase, have even higher rates of mass loss.[4]

The middle-aged Sun

According to the astronomer Richard Radick,[5] when the Sun entered middle age 2,500 Myr ago gone was the intemperance of youth. In came a dignified mode of existence, with the need for an occasional nap, a state which, to judge from its fellow stars, is likely to persist for the remaining 4,000 Myr of the Sun's main-sequence lifetime, although its rate of rotation and level of activity will slow down further. Among sunlike stars middle age lasts a very long time.

Younger main sequence stars do indeed behave erratically and vary in luminosity by several percent, whereas most middle-aged stars vary only by about 0.1% and display some kind of regular cyclicity, often with a period of about a decade. Regrettably the Sun's record on this score is too short for any kind of confident assessment. Dependable values for its luminosity were first obtained by orbiting satellites in 1978. Even if we include ground-based data, with all the errors introduced by the atmosphere, the period of record did not begin until the bolometer readings on Mt Whitney in 1881. For earlier times we have to rely on what are called proxy data, a misleading term because, in legal circles at least, proxies are meant to be dependable whereas many solar proxies are of dubious character and we have little way of knowing how much faith they should inspire.

In meteorites, the isotope of titanium ^{44}Ti, as we saw earlier, can register GCR activity in space free from any interference from the Earth so that the solar component of any contemporaneous records on Earth can be corrected for GCR flux. Neon 21 (^{21}Ne) provides a direct response to intense solar flare activity on the Sun and as it has been recovered from the oldest meteorites known it reaches back to 4.5 Gyr ago[6] but here too we are currently limited to a few spot measurements.

The Moon has been described as a giant tape recorder for solar (and other solar system) events.[7] Since the 1970s analysis of deposits on its surface which have been preserved by burial and sampled by astronauts has made it possible to trace variations in the solar wind and the incidence of solar flares. Although exposure times appear to range between 1,000 and 10,000 years, the record here (as noted earlier) appears to go back at least 1 Gyr. The solar wind implants a variety of elements in the Moon's surface soil, or lunar regolith, three times as effectively on the far side as on the nearside, where it is shielded by the Earth's magnetic field for part of the solar orbit. It does so only to a depth of less than 1 micron (μm), but meteorite impacts and gravity conspire to stir the surface layers into the soil rather like our earthworms.[8] Even so, as with meteorites we are some way from securing anything approaching a sequence of measurements.

There is an interesting twist to the lunar evidence. One of the implanted components of the solar wind is helium 3 (^{3}He), a potential fuel for clean-burning fusion reactors but at the moment of value solely for fusion research. According to its proponents, who including the lunar astronaut Harrison Schmitt, the energy yield of ^{3}He when it is burned with deuterium is about 300:1 (about 44:1 if you allow for the energy used to build the plant and to make the fusion take place), compared with 16:1 for coal and 20:1 for nuclear fission. There is very little radioactive waste.

In testimony to the USA House Committee on Science and Technology in April 2004, it was reported that, on the basis of the samples brought back by Apollo astronauts, who drilled into the Moon to a maximum depth of 3 m, there are between 450,000 and 4.6 million tons of ^{3}He in the regolith. But the concentration is 20–30 parts per billion, so that the 25 t estimated as sufficient to meet current US energy needs for a year would require the processing of $1,250 \times 10^6$ t of regolith. Of course, if the extraction is done on the Moon the ^{3}He can be used both for fuel in lunar bases and for export to Earth.

Politics inevitably entered the picture. The ^{3}He provides economic justification for lunar bases, and one might add studies of solar physics. The threat of an international scramble for this potentially valuable resource adds an element of urgency to the analysis. In 2006 Russia's Energia Space Corporation announced plans for a ^{3}He mining base on the Moon, and China launched its first lunar probe, Chang'e I, in October 2007 with the avowed aim of mapping its lithosphere and in the belief that 'each year three space shuttle missions could bring enough fuel for all human beings across the world'.[9]

The helium is trapped most efficiently by ilmenite (iron titanium oxide), whose distribution can be detected from lunar orbit and which can also yield the oxygen required for lunar bases. Information on the mechanics of the solar wind will of course help in estimating the reserves of ^{3}He and other potential resources both outside the area explored by the Apollo teams and at depths greater than 3 m. In other words 'academic' studies of changes in the vigour of the Sun have practical implications.

How are those changes reconstructed? The events of the last half-million years of solar history are documented by the cosmogenic isotopes ^{18}O and ^{14}C which cannot be used as measures of the strength of the solar wind without some information on the strength of the Earth's magnetic field at the time in question. There are two potential sources of information on this score which illustrate anew the wonderful (or irritating) range of material that needs to be exploited in solar studies: ancient kilns, which clearly do not go back in time much before classical Antiquity, and volcanic rocks, which do but are often difficult and expensive to reach because they are on the sea floor or deep underground. Why old kilns and lavas? Because when rock or pottery clay is heated above a critical temperature called the Curie point it memorises the direction and to some extent the strength of the local magnetic field and retains these items once it has cooled down always providing of course that it is not moved from its original position.

The results of many measurements on kilns (or on pots still in them) and on lavas in many parts of the world are widely seen as evidence that changes in the geomagnetic shield could account for fluctuations in ^{14}C lasting a few millennia whereas the solar factor can explain changes spanning a few decades or centuries (Fig. 3.1; see also 3.5). But the geomagnetic field record is very patchy and not at all consistent across the globe, and big swings in the GCR flux over the millennia cannot yet be ruled out.[10]

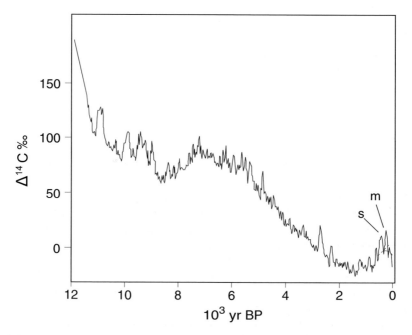

Fig. 3.1 The radiocarbon (^{14}C) content of tree rings for the last 12,000 years is a mirror image of solar activity because the galactic cosmic rays (GCRs) that create ^{14}C in the upper atmosphere are deflected by the solar wind. Unfortunately the picture is complicated by changes in the Earth's magnetic field and in climate. s = Spörer Minimum, m = Maunder Minimum

To resolve the matter – and it is worth resolving because it could indicate substantial swings in the Sun's output in the recent past and conceivably therefore in the near future – we must turn to ^{10}Be, the other cosmogenic isotope in the frame. ^{14}C may loiter in the atmosphere for years and thus be wafted far and wide before it is plucked out and stored by any living thing which breathes air, depends on photosynthesis or eats plants that do; a ^{14}C sample from any single location therefore bears on conditions in many parts of the world. ^{10}Be is washed by rain or snow out of the atmosphere a year or two at most after it formed and is trapped in lake muds, ocean sediments, snow or ice; its impressions are therefore relatively parochial.

More important still in the present debate is that ^{10}Be levels at high latitudes do not appear to be greatly influenced by changes in the geomagnetic field.[11] As the half-life of ^{10}Be is 1.51 million years, compared to the 5,730 years for ^{14}C, there is every chance that ^{10}Be production will be a good measure of solar activity for much longer than the 50,000 or so years documented by ^{14}C.

The fine detail of the last few centuries or so of solar history comes from sunspots backed up by auroras. The agreement between them is good, hardly surprising as they both respond to, or signal, solar activity. The question is whether longer-term and possibly more violent fluctuations in the Sun's output are being overlooked purely by default.

Sunspots and auroras

The earliest observations of sunspots through the telescope made it obvious that they moved but the Sun's flawless complexion could be safeguarded by dismissing the spots as orbiting planets. Galileo scotched this idea by showing that they changed shape and size as they moved and he suggested that they were analogous to terrestrial clouds. Both Galileo and Fabricius concluded that the Sun rotated and Galileo put the period of rotation at about 1 lunar month. He also noted that the spots moved relative to each other but were largely confined to two belts adjoining the equator.

Scheiner, a Jesuit who was not allowed by his superiors to publish his sunspot work in his own name, was one of those who thought the spots were planets. He was soon put right by Galileo in a series of letters. In 1630 Scheiner published the fruit of several years of observation, the *Rosa Ursina, sive Sol*, dedicated to the Duke of Orsini whose family crest included a rose and a bear. In it he accepted that the sunspots were on the Sun's surface; more important, he showed the path of spots across the solar surface (Fig. 3.2). His careful drawings have also proved invaluable to studies of sunspot frequency in the past.[12]

It still took until 1843 for Samuel Heinrich Schwabe, a pharmacist, to realize that the pattern of the sunspots followed some kind of period. His first estimate, after 17 years of observation, was 10 years. In 1848 Rudolf Wolf announced an improved estimate of 11 years (the range is now known to be from 9 to 13.6 years

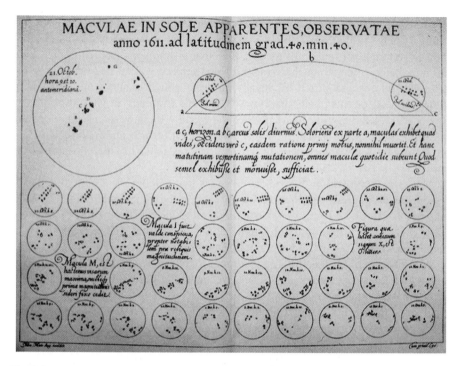

Fig. 3.2 Some of Christopher Scheiner's sunspot drawings for 1611. Note change in the latitude of sunspot groups

although since 1848 the range has narrowed to 10–12 years) and he introduced the relative sunspot number R which remains the accepted measure of sunspot activity and which takes into account the total number of spots and the number of spot groups on the sun's disk. According to this scheme R is defined as K(10g +f) where f is the number of spots, g the number of sunspot groups and K a personal coefficient which allows for counting method, size of telescope, viewing conditions and what has been called 'observer enthusiasm'.[13] This sounds hopelessly vague but it can be standardized for individual observatories. Thus K at Zurich Observatory has been maintained at 0.6 since 1882.[14] The amplitude of the solar cycle is also very variable, with annual mean spot numbers[15] usually ranging from 45 to 190. The lifespan of any one sunspot or sunspot group is usually less than 2 days but some spots last several months.

In 1858 another amateur observer, Richard Carrington, reported that the spots moved closer to the solar equator in the course of the solar cycle. Although Scheiner had noted a change in latitude, the latitude effect came to be known as Spörer's Law after Gustav Spörer, who added to Carrington's observations. It was to be immortalized in the butterfly diagram first drafted by W.A. Maunder, which shows the progressive equatorward drift from latitudes 20–35° and the general increase in sunspot size as the cycle progresses (Fig. 3.3). Carrington also noted that the rate

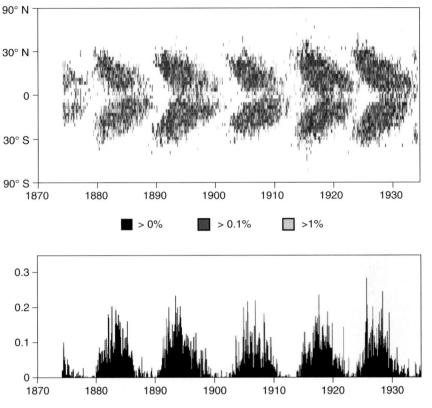

Fig. 3.3 Butterfly diagram. The upper figure shows sunspot area as percentage of latitude, the lower figure shows average daily sunspot area as percentage of the visible hemisphere (Courtesy of NASA)

of rotation indicated by sunspots varied between the high and middle latitudes. A sunspot group at the Equator nowadays completes a full rotation in 26.8 Earth days and one at 60°N takes 30.8 days,[16] a difference of 13%. Differential rotation is of course central to the Babcock model of sunspot origin.

The Sun's magnetic field, like the Earth's, is broadly aligned with the axis of rotation and it reverses from time to time, although Earth reversals are separated by tens or thousands of years whereas reversals on the Sun occur roughly every 11 years. Some sunspots are in pairs, with the sunspot that is leading in the direction of solar rotation magnetically positive and the following one negative as for a bar magnet. In the other hemisphere the situation is the reverse (Fig. 3.4). The sunspots first appear at high latitudes, gradually migrate towards the equator and eventually disappear, leaving a field in which the new, migrating pairs are of the opposite polarity. The progress from high to low latitudes coincides with the rise and fall in sunspot number – the solar (or Schwabe) cycle. The magnetic cycle, where the sunspot polarity reverts to its initial state, is therefore twice as long and is named after Hale, the discoverer of sunspot magnetism.

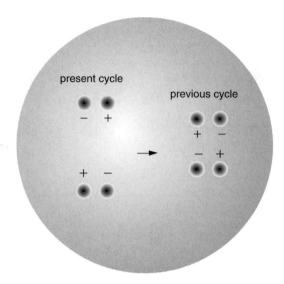

Fig. 3.4 The Hale cycle refers to the reversal of the Sun's magnetic field during every 11-year sunspot cycle so that it returns to its original state after about 22 years. The same applies to the magnetic polarity of sunspot pairs

Spörer had noted in 1887 that sunspots were few between AD 1645 and 1715. In 1922 Maunder, who was in charge of the solar division at the Greenwich Observatory, used its archives to investigate the question. His announcement of 'a prolonged sunspot minimum' was first largely ignored and later greeted with scepticism, as any fall in the number of recorded sunspots was commonly blamed on a lack of methodical observers. The case was argued with special vigour when it came to the Chinese evidence. Unless the observer leaves proof that he or she was observing throughout the period in question and the proof survives intact (it was suggested) a nil result remains ambiguous. Histories of the Ch'ing dynasty provided a special case for, unlike those of its predecessors, they just omitted sunspot chronologies, and any that slipped through could well have been destroyed during the Boxer Rebellion, when Allied troops occupied the Peking Observatory.[17]

The evidence for a Maunder minimum in the European records is more difficult to dismiss in this fashion especially as the astronomers of the time, who had been alerted to sunspots by Galileo's observations 35 years earlier, are likely to have recorded them with scrupulous care precisely because they were rare. John Eddy, a solar astronomer at the High Altitude Observatory in Boulder, took up the challenge, having initially set out to confirm that the Sun's output was steady. He based his analysis (Fig. 3.5) on observations made in Germany in 1652–1685, France in 1653–1718, England in 1676–1699 and more spasmodically elsewhere, and he found that between 1672 and 1699 fewer than 50 sunspots were reported when in the 20th century a typical 30 year period would have seen 40,000–50,000 spots.[18]

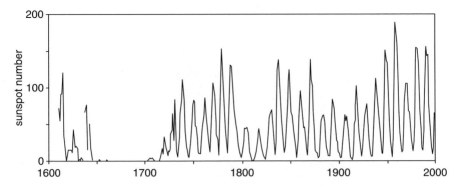

Fig. 3.5 Sunspot number graph showing lack of sunspots between about 1630 and 1710 – the Maunder Minimum. Supporting evidence has come from aurora reports and the ^{14}C record

The time taken by a single solar rotation appears to have risen from a few days when the Sun was 100 Myr old to perhaps a week at 700 Myr.[19] Thanks in part to the drag of the solar wind the Sun's spin rate has continued to decline to its present level. Comparison with sunspot drawings made in 1620 shows that the rotation rate then was much the same as today's whereas data from the period of the Maunder Minimum suggests that the equatorial rate was then significantly faster at nearly 14° per day compared to the present 13°. If anything this is not consistent with the Babcock model for sunspots, as we might expect more strangulation of magnetic field lines to go with faster spin rates, but the answer may lie in the level of differential rotation – a facet of the record which will be difficult to reconstruct.

Eddy found further confirmation for the Maunder Minimum in the radiocarbon incorporated in tree rings, whose annual growth rings provide unusually precise dating (Fig. 3.1). Ring sequences can be obtained from long-lived trees such as the Bristlecone pine (*Pinus longaeva*), of which an individual over 4,600 years old has been found living in the White Mountains of California, the Douglas fir, the North Pacific sequoia and a number of European oaks. By combining their ring sequences with those derived from timbers found in old buildings or archaeological sites which overlap in age one can construct composite sequences which can then be used to trace changes in the atmospheric content of ^{14}C as well as local climate and ground conditions.

The period between 1650 and 1750 shows a marked increase in ^{14}C levels consistent with a weakened solar wind. At least 12 other solar minima have been inferred from the ^{14}C record of the last 5,000 years. Some of them are deep and long enough to acquire a name: the Oort (AD 1010–1050), the Wolf (AD 1280–1340), the Spörer (AD 1420–1530) and the Dalton (AD 1790–1820). Eddy also identified a number of positive solar anomalies during that interval, including the Medieval Maximum, which came just before the Spörer, the Stonehenge Maximum, the Pyramid Maximum and the Sumerian Maximum. The names derive from simple age rather than to cultural connotations but Eddy did not hesitate to link the most recent ^{14}C events with glacial history and temperatures in Europe.[20]

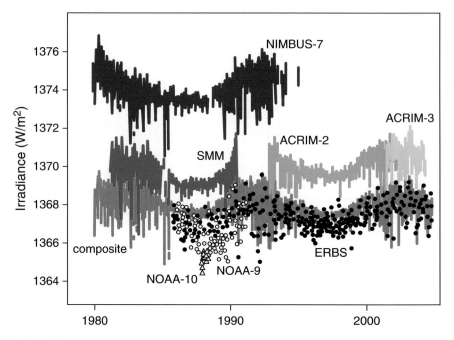

Fig. 3.6 Changes in solar output measured by satellite during two solar cycles. The absolute values vary between the instruments but the cycles are clearly represented. The ERBE instruments were launched by the Space Shuttle Challenger in 1984 on Earth Radiation Budget Satellite (ERBS) and on two National Oceanic and Atmospheric Administration weather monitoring satellites, NOAA 9 and NOAA 10, in 1984 and 1986

As the processes behind sunspots are not fully understood, and as their effect on solar luminosity is outweighed by faculae, sunspots remain a convenient but vague guide to the level of solar activity. Nevertheless satellite data over three solar cycles since 1978 have confirmed that the Sun is brightest when its face is spottiest. The claim would seem not to amount to much when we consider that measurements during 1978–1993 from the Nimbus-7 satellite showed that the radiation from the Sun at solar maximum was only 0.2% more than at solar minimum. The Solar Maximum Mission satellite (1980–1989) also showed little difference between solar maximum and solar minimum (Fig. 3.6). But these numbers conceal a crucial detail: not all parts of the solar spectrum change by the same amount during the sunspot cycle. The UV part (100–400 nm), which contributes only 1% of the Sun's total output, strongly affects the heating, composition and circulation of the Earth's middle and upper atmosphere and it governs the Sun's effect on human health. To judge from the period 1980–1986, radiation at 10–120 nm, which includes part of the UV component, can be as much as 10 times larger at maximum than at minimum, and X-rays (<10 nm) as much as 1,000 times larger.[21]

The ^{14}C curve has since been extended well beyond 3000 BC using overlapping tree-ring sequences and, when these run out, corals dated by other chemical methods. Such composite curves are widely used to correct dates for the last 10,000 years which are based on the radiocarbon method as they show how far the age indicated by ^{14}C analysis diverges from true (calendar) age.[22] They also show that the ^{14}C blips matched by Eddy to sunspot minima and maxima ride on a broad series of undulations which peaked a few centuries ago and also about 6,000 years before that. The wave rises again somewhere over 12,000 years ago but we then run out of dependable atmospheric data.

The 11-year sunspot cycle and its 22-year magnetic cousin, the evidence for several historical minima, and the hint of grander oscillations raise the crucial question of whether there is some kind of timekeeper in the Sun. Any evidence of genuine cyclicity if of course critical to attempts at prediction; but it is just as important to students of the physics of stars. Besides the approximate 11-year cycle and the 5,000 undulation in the ^{14}C curve there is some evidence of a 70–90 year periodicity (the Gleissberg cycle) and a stronger 210 years one (the de Vries cycle). There is also the hint of a 2,200–2,400 year pulse during the last 8,000 years.[23]

The sequence of the five pronounced minima of recent centuries – the Oort, Wolf, Spörer, Maunder and Dalton – may in fact reflect the de Vries cycle,[24] the implication being that a minimum is to come in about AD 2050, but measurements from satellites are thought to indicate the opposite. Six overlapping satellite experiments have measured total solar irradiance since late in 1978; the minima of Cycles 21–23 show a significant positive trend of 0.05% per decade.[25] This is a small amount and could soon be cancelled out but it illustrates the refined nature of satellite observation of the Sun. More to the present point, as a 0.25% fall in solar luminosity is thought to have driven the Little Ice Age, 5 years at −0.05% should perhaps be enough to start a new one.

Support for the reality of the sunspot minima and their correspondence with a weakened solar wind has come from aurora observations, which require no equipment and can be made by people in all walks of life. Maunder himself noticed that few auroras were reported during 1645–1715. In 1890, when he was writing, the existence of a solar wind was not suspected and the link between sunspots and colourful skies was wholly puzzling, not least because, whereas changes in the photosphere are observed eight minutes after they occur, the solar wind particles take about 80 h to make the journey to Earth.[26]

Auroras include mixtures of colours. Green, white and red predominate in auroras on Earth because charged particles in the solar wind collide with and excite oxygen and nitrogen in the atmosphere. Oxygen emits photons at green wavelengths (558 and 630 nm). It also emits pure red wavelengths (630 and 636 nm) at high altitudes, where it responds to slower particles, and a reddish tinge at the base of some green auroras when it interacts with particles at relatively low altitude. Nitrogen itself emits at blue (428 and 391 nm) but human vision is poor in the blue wavelengths, so that green, especially the yellowish variety at 558 nm, thus tends to dominate.[27]

Nevertheless the link between sunspot minima and a lack of auroras is not straightforward, as it is affected by cultural and historical factors perhaps more markedly than sunspot observation. Very few auroras were reported even before 1645, although numbers had increased substantially after 1550 thanks to the growing interest in astronomy fostered by the Renaissance. Again, their recovery after

1716 is of course consistent with the end of a solar minimum but owed something to an influential paper by Edmond Halley, the Astronomer Royal, which sought to explain the novel phenomenon.[28]

Bearing these limitations in mind, the auroral record provides strong support for the link between sunspots and the energy of the solar wind. Detailed analysis of about 45,000 observations made during 1450–1948 showed that the incidence of auroras[29] was low during the Dalton, Maunder and Spörer as well as at around 1765. Other, weaker correlations between sunspot series and auroras have been noted.

Shortlived events

The drift and changing shape of sunspots were leisurely effects which could be caught by regular but not necessarily continuous observation. Shortlived events were revealed only by accident. The most memorable took place on 1 September 1859, when Carrington observed a violent flare on the Sun. He had been making his daily sunspot observation when 'two patches of intensely bright and white light broke out'. By the time Carrington had found someone to confirm his observation the flare had faded. Even so he had time to observe that the phenomenon took place at an elevation considerably above the general surface of the sun, and, accordingly, altogether above and over the great [sunspot] group in which it was seen projected.[30]

The 1859 flare was accompanied by serious disruption of telegraphic links as well as extensive auroral displays over much of Europe, events which dramatically revealed the capricious and violent nature of the Sun.

Flares are on the whole more common near sunspot maxima than at other times, although two notable exceptions are noted in Chapter 7; they are electromagnetic explosions in the corona which take place when magnetic fields pointing in opposite directions suddenly interact and coronal temperatures of a few million degrees are raised to as much as 100 million degrees.[31] Lasting 1–2 hr (or much less, as in Carrington's account) they emit UV and visible light and X-rays as well as protons and electrons at speeds close to 1,000 km/s. Flares are classified according to their X-ray brightness (at 10–80 nm and measured in watts per square metre) in three classes each of which has nine subdivisions: X-class, which can trigger radio blackouts across the Earth and launch radiation storms; M-class, which may cause brief radio blackouts in the polar regions, and C-class, which have little effect on Earth.

Flares increase the ionization in the Earth's ionosphere, hence some of the problems with radio reception discussed in Chapter 7 they may also generate solar 'quakes which, as shown in Fig. 3.7, are manifested as surface ripples moving up and down on the Sun's photosphere. The ripples are due to standing acoustic waves trapped between regions of the solar interior of different densities,[32] and have a period of roughly 5 min and speeds of 500 m/s.

Before 1961, all we had in the way of data on the changing Sun was information on its radius, luminosity, mass, composition and age, and the last two were, apart from surface spectroscopic measurements, a matter of inference. Besides values for

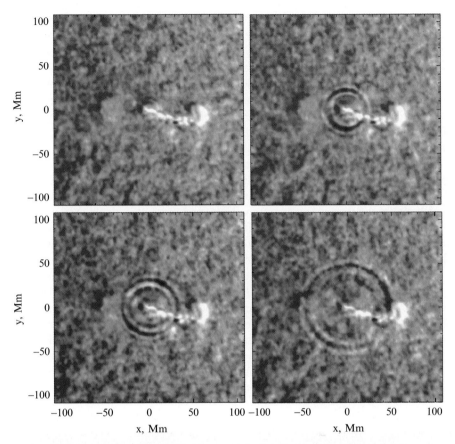

Fig. 3.7 Surface ripples resulting from a sunquake caused by a flare on 9 July 1996. The magnitude of the quake using a terrestrial scale is 11.3, 40,000 times the energy released in the San Francisco earthquake of 1906. Data from Doppler imager on SOHO (Courtesy of AG Kosovichev (Stanford) et al., MDI, SOHO, ESA and NASA). The ripples are about 2 km high

helium content and for the depth of the convection zone helioseismology has provided a clear indication that the convective and radiative zones rotate at different rates, making the tachocline that separates them the likely source of the Sun's magnetic field. Helioseismology has also yielded estimates for temperature at different depths within the Sun and shown that the convective zone is disrupted by horizontal jet streams.

But traditional sources have not shot their bolt. Flares, as Carrington shrewdly noted, often come from above active regions on the Sun which are rich in sunspots. So do coronal mass ejections (CMEs) although they are no longer thought to be caused by flares. CMEs occur on average 15–20 times a week near solar maximum and once a week near solar minimum. As a CME travels through the slower solar wind a shock front may form before it which on meeting the Earth's magnetic field produces auroras much nearer the Equator than usual.

Granted that the amount of matter discarded as CMEs is a trivial portion of the Sun's total mass, they bring us to the question of its changing dimensions. Some workers think that oscillations lasting 80 years may affect the solar constant.[33] Measuring any such effect is of course challenging. One method uses the transit of the planet Mercury: comparison of the time it took in 1715 with that in 1979 suggests that the Sun's diameter has fallen by an average of just over 11 km/yr. A similar result is obtained by comparing the width of annular eclipses at successive dates. There is also some evidence that the Sun had a larger diameter as well as rotating more slowly during the Maunder Minimum.[34]

In view of the difficulties faced by human observation of these phenomena, however, and the fact that theory predicts a reduction in the Sun's diameter of only 2.4 cm/yr, the sceptics have turned to electronic substitutes, including a device which makes the measurement automatically and more objectively. Yet even their most devoted champions admit that it will take several million years of measurement for a confident conclusion. On the other hand the SOHO satellite has shown that the total irradiance of the Sun has been increasing steadily since 1996.

Other kinds of imaging reveal the dynamic nature of what are called active regions around bipolar sunspot pairs. At extreme UV with a wavelength of 17.1 nm, magnetic loops of ionized gas or plasma can be seen rising up to 500,000 km above the photosphere and linking two sunspots (Figs. 1.9d and 3.9). Prominences are regions of dense gas kept aloft above the photosphere by magnetic fields, and include quiescent (long-lived) and active (short-lived) varieties; the latter are associated with active regions (Fig. 3.8). Filaments, which appear as prominences when viewed on the Sun's limb, are also controlled in their shape and development by

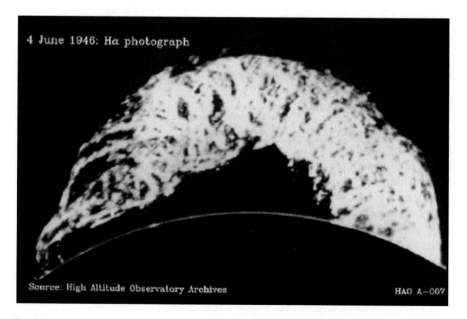

4 June 1946: Hα photograph

Source: High Altitude Observatory Archives   HAO A–007

Fig. 3.8 Prominence imaged in H α light in 1946 (Courtesy of NASA)

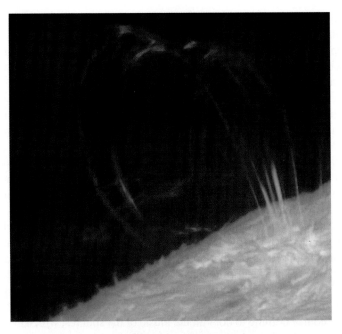

Fig. 3.9 Coronal loops formed after an active region on the Sun flared 26 June 1992. Material moving away from the observer is shifted towards the red end of the spectrum, and if moving towards the observer it is shifted towards the blue (Courtesy of NASA. Images from Swedish Solar Telescope on La Palma)

local magnetic fields (Fig. 1.9b). The granules on the photosphere have lifetimes of 3–10 min (Fig. 2.2) and the supergranules may last several days.

The bonds that bind Earth and Sun include the interplanetary magnetic field, nourished by the solar wind, and the geomagnetic field. Data obtained by the Ulysses spacecraft since 1990 show that the solar wind is equally strong at all solar latitudes, which means that historical data from the Earth can be used to investigate the Sun's magnetic history. The impact of the Sun on the Earth's magnetic field has been measured on the ground since 1868 using magnetometers in England and Australia. Not only does the Earth's field respond to the 11 and 22-year cycles; the total coronal magnetic field leaving the Sun[35] has reportedly doubled since 1901. The search is now on for evidence of corresponding changes in stratospheric chemistry and in cloud cover so that the solar contribution to recent and future climatic change can be evaluated objectively.

A few years ago when preparing to give a talk on *The Inconstant Sun* to a mixed audience I wondered how to convey the notion that the 11-year sunspot cycle is probably superimposed on oscillations with periods measured in centuries which in turn ride on others spanning millennia. Rather than draw a whole lot of curves on the board I decided to make the point using sound, and lit upon Bach's prelude no 6 in E (Fig. 3.10).

This is of course nothing to do with the music of the spheres, which even the great Kepler, using standard musical notation, identified in the motion of the planets.[36] Many performances of the prelude are so slow that the repeated triplets blot out

Fig. 3.10 JS Bach, Prelude no 6. Three different melodic 'wavelengths' (I + II, III + IV, and V) are highlighted

everything else, but one that is fast enough and with the right emphasis will bring out at least three 'wavelengths', the shortest being the triplets, the second (as I have tried to show in the figure) composed of pairs of phrases, and the third a melodic carrier wave which conveys the argument of the prelude or what is sometimes termed the contour of a piece. To caricature the analogy, the triplets could be said to represent the 11-year solar cycles and the pairs the 22-year magnetic cycles. The third period is then perhaps the elusive wavelength of the radiocarbon record. Such patterns emerge only when the sequence is long enough and is played distinctly. And, as in Bach's music, in Nature you have some idea of what to expect but are usually caught out.

Chapter 4
Sun and Climate

The Sun supplies 99.998% of the energy that drives the atmosphere; the remaining 0.002% comes from the Earth's interior. Yet, although we know perfectly well that the climate gets hotter as we approach the Equator and colder the other way, and that we have seasons because the Earth is tilted relative to the Sun so that it is overhead in summer and not in winter, there is wide disagreement about the climatic impact of *changes* in the Sun.

To the newcomer it seems difficult to believe that there should be any uncertainty about the matter. But in the words of Eugene Parker,[1] discoverer of the solar wind, the biggest mistake that we could make would be to think that we know the answers when we do not.

Parker was referring to assumptions about the significance of the solar factor which are based on hunches or less. A review published in 2004 baldly stated that the length of modern instrumental records is *still too short to unambiguously identify a solar influence on climate*.[2] Even if we disregard the invention of the barometer in 1643 and only consider measurements of temperature and rainfall which have been made after the telegraph allowed records to be collated worldwide, that claim amounts to dismissing over 130 years of painstaking data gathering. How can that be?

There are at least three reasons. First, our grasp of solar history, as we have just seen, is still very shaky, and the instrumental bit of it satisfactorily covers a few decades at most. It is difficult to analyse climatic change when trends, let alone cycles, are difficult to establish securely.

Second, the physics and chemistry of any solar influence on climate is either poorly understood or highly controversial. Is it simply a matter of heating or cooling of the atmosphere? Is the ozone content of the stratosphere critical? Or do cosmic rays hold the key, whereupon the Sun's role is mainly to act as gatekeeper?

And third, the climate machine is very complex. The net radiation received by the Earth as a whole is governed not only by fluctuations in the Sun's output but also by the shape of the Earth's orbit and the alignment of its axis, all of which change over the millennia. Almost a third of the radiation received by the Earth is immediately reflected back into space (Fig. 4.1). The reminder is absorbed by the atmosphere, the land and the ocean. The last two return that energy to the atmosphere mainly as IR radiation some of which generates water vapour which releases heat when it condenses as rain or clouds.

C. Vita-Finzi, *The Sun: A User's Manual*,
doi: 10.1007/978-1-4020-6881-2_4, © Springer Science+Business Media B.V. 2008

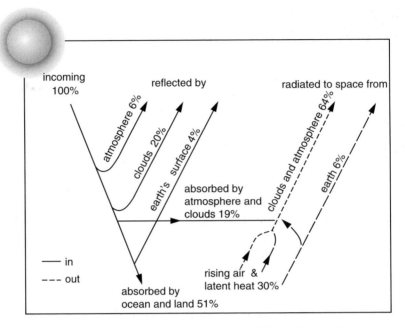

Fig. 4.1 The Earth's radiation budget today. It represents the balance between incoming energy from the Sun and outgoing thermal (long wave) and reflected (short wave) energy from the Earth (Courtesy of NASA)

Moreover, the atmosphere is constantly on the move and its composition changes from place to place and from one moment to the next; the cloud cover reflects the incoming solar energy (the percentage reflected being known as the albedo[3] as well as trapping some of the energy radiated back by the Earth. The distribution of land and sea, of topography and of vegetation, is also constantly changing with the seasons, fluctuating sea level, and the vagaries of erosion by water and wind. Many of these factors respond to positive or negative feedback and sometimes both. As an instance of the former, when snow accumulates its high albedo reduces the amount of solar radiation absorbed by the ground and the snow persists; if for some reason it starts melting and turns slushy and grubby it absorbs an increasing amount of solar radiation and melts all the faster. Negative feedback prevails when (for example) evaporation from the sea on a warm day generates clouds which then blot out some of the sunshine.

Many of the key discoveries in weather analysis, the recognition of high and low pressure systems, air masses and fronts, of the global circulation, and of the mechanisms by which rain and snow are produced, date only from the early decades of the 20th century. They provided the mobile scaffolding on which to hang meteorological data, and thus to trace the atmosphere's three-dimensional evolution from hour to hour. A number of the most influential pioneers practised their craft in the pale, high latitudes of Scandinavia, where there are times when Sun either never sets or refuses to rise. This was possibly an advantage when it came to recognising the important role in the development of weather played by the displacement and interaction of homogeneous parcels of air. But, doubtless in part because there were still no direct measurements of solar radiation by satellite-borne instruments, the Sun's influence was given little prominence.

For all these reasons the solar factor long remained elusive. In 1945 the mathematician John von Neumann was looking for a scientific problem which was difficult enough to demonstrate the potential of a stored-program computer. He settled on weather prediction.[4] It is claimed that von Neumann was not motivated solely by scientific curiosity and that, as a staunch anti-Communist, he hoped the results of weather analysis would lead to weather control so that one might ruin Soviet harvests.[5] Stalin, aided by the Sun, managed perfectly well without von Neumann's help; in Moldova, for example, famine followed the requisitioning of large amounts of agricultural products which was carried through even though a catastrophic drought in 1945–1947 had precipitated a poor harvest. Whatever his motives, von Neumann made little headway.

The faint young Sun

Von Neumann began his analysis with a uniform rotating sphere on which the amount of solar radiation depended on latitude but was not influenced by longitude. Many accounts of the Earth's early climate are forced to be just as sketchy by our inadequate grasp of past geographies and topographies.

If we accept that the Sun is indeed progressing along the Main Sequence (Fig. 2.8), the average temperatures of the early Earth should have been low enough to leave any surface waters frozen. Yet there were liquid oceans on Earth at least 4.35 Gyr ago.

This is the *faint young Sun paradox.*

The explanation that has gained the greatest support depends on the greenhouse effect, a notion which has become familiar in connection with the problem of global warming today. The glass of a garden greenhouse lets in sunlight (typically 280–2,500 nm) and traps the outgoing IR radiation (5,000–35,000 nm) mainly by inhibiting convection. Certain atmospheric gases, notably CO_2 and water vapour, let through long-wave radiation, absorbing the outgoing IR radiation, and reradiating it back to Earth. The greenhouse analogy is therefore mainly figurative.

The faint young Sun paradox can be resolved by assuming that the early atmosphere was richer in greenhouse gases than it is now. The first gas that was proposed for this role was ammonia as it is very effective at trapping radiation. But it is easily destroyed by solar UV radiation if the atmosphere does not contain oxygen too. The next candidate was carbon dioxide (CO_2). Estimates for the amount of this gas that would have been required to resolve the paradox are in the region of 50 times the present level, a quantity that could have been supplied by volcanic eruptions but far exceeds what is indicated by the geological evidence.

If supplemented by an atmospheric content of 0.1% of methane, however, the CO_2 levels that seem plausible would have kept global temperatures above freezing. Methane is more effective than CO_2 because it absorbs a wider range of wavelengths;

and it could have been supplied by some of the earliest microbes.[6] A rise in global temperatures would have favoured the growth of the methane-producing bacteria, the methanogens. After a while, however, negative feedback may have come into play as the methane in the atmosphere formed a hydrocarbon haze which absorbed some of the incoming sunlight and reradiated it into space.

There is thus no need to have the Sun break ranks by cooling rather than warming as it matured. Yet that solution to the paradox has its attractions. Among other things it would help to resolve the puzzling evidence for a warmer, wetter past on Mars. And an increase in luminosity of 5% can be obtained by raising the mass of the Sun by as little as 0.07%, with the added advantage that the enhanced gravity would have drawn the planets closer to the warming Sun.

But it may just be that the Earth was largely oceanic during its first 1,500 Myr of its existence[8] and, having a correspondingly lower albedo, it absorbed more of the incoming radiation than if it had been composed predominantly of dry land. Indeed, the oceans may have been hotter because there was more heat coming from the interior of the Earth. Consequently they gave off more water vapour and thus stoked up the greenhouse effect.

The Sun would have added to the confusion by breaking up any oxygen (O_2) in the stratosphere into its constituent atoms (O) which then combined with any spare O_2 molecules to make ozone (O_3). Ozone absorbs solar radiation and thus warms its surroundings, an effect which, even if trivial in these early times, was to acquire great significance 4.5 Gyr later, during what we call the Middle Ages.

As (or if) the Sun warmed up, the greenhouse effect had to be reduced if the Earth was not to suffer the fate of the planet Venus and experience a runaway greenhouse, with most of its water in a thick cloud cover, temperatures of 480°C. and a surface pressure equivalent to 90 Earth atmospheres. For whatever reason the early high levels of carbon dioxide in the atmosphere fell and the gas was entombed in the calcium carbonate of coral, mollusc shells and the tiny foraminifers that make up the Chalk. Today 99.9% of the world's CO_2 takes the form of carbonate rocks.

There is some dispute about why this came about. According to the Gaia hypothesis living organisms manipulated atmospheric composition in order to stabilize surface temperatures at a level which they found favourable. Austere evolutionary biologists will prefer a narrative in which organisms simply adapted to the changing conditions, with the rate of CO_2 fixation initially high and falling as temperatures declined. In one scheme all is clever tuning, in the other, life muddles through.

Ice ages

The Earth's receipt of solar energy before anything in the way of a greenhouse comes into play is evidently influenced by the geometrical arrangement in space of the two bodies. By 320 BC Greek astronomers were aware that the Earth's axis of rotation describes a cone roughly every 6,000 years; the Pole Star, for example, now in the Little Dipper, was Thuban, or α Draconis, in the days of Pharaonic Egypt and it will revert to being so in due course.

But the Ancients had no reason to suspect that our distance to the Sun was subject to change. Once again a notion due to Aristotle put a brake on progress for a couple of millennia. He maintained that Nature favoured orderly forms and that the various bodies in the solar system therefore followed circular paths. Even the arch-rebel Galileo Galilei plumped for a circular orbit for Earth, as did the father of the Sun-centered model of the heavens Copernicus.

Not so Kepler, who realized from the observational data amassed by his predecessors that the planets followed elliptical orbits. It followed that they were not at a constant distance from the Sun. An AU is defined as about 1.5×10^8 km, but this is an average. The distance between Sun and Earth varies by a little over 3% during the year, equivalent to 7% in the level of radiation from the Sun.

The ellipticity or eccentricity itself ranges between 0.005 and 0.06 (it is now 0.017), and it reverts to its original value every 105,000 years. This is unlikely to have much of a direct effect on climatic history. But it may amplify the role of precession. Precession determines the time of year when the Earth is closest to the Sun (perihelion) and gradually alters it over a 21,000 year cycle. When the eccentricity is high (never very high for Earth, but at maximum it is three times greater than at minimum) the effect of precession is more pronounced.

There is a third source of long-term change: the angle the Earth's axial tilt makes with the plane of its orbit, hence its tonguetwisting name of 'obliquity of the ecliptic'. It is currently 23.5° and it ranges between 22.1° and 24.5°, with a period of 41,000 years. A more pronounced tilt should lead to increased contrast between the seasons.

The three factors, eccentricity, precession and axial tilt (Figs. 4.2a, b) are named collectively after Milutin Milankovitch, a Serbian astronomer who in 1910 formalised their analysis. But they were known to earlier workers. At a time when most geologists explained glaciation by changes in topography and oceanic circulation and did not generally accept more than one ice age, James Croll[9] argued for astronomical mechanisms. For instance, a very elongated orbit at a time when the winter solstice occurred far from the Sun could precipitate an ice age. It followed that there had been numerous glaciations in one or other hemisphere, and Croll was well aware of positive feedback, such as radiation from ice fields. Alfred Russel Wallace, co-discoverer with Darwin of evolution by natural selection, went so far as to suggest that Croll's explanation also applied to Mars, and in 1881 he reported evidence for the effects of precession at a time of moderate eccentricity in the advance and retreat of the Martian ice caps that could be seen through telescopes.

Milankovitch developed Croll's ideas, notably by working out mathematically how insolation would vary with latitude for the last 600,000 years. General acceptance of his model was long delayed. A major obstacle was the ingrained belief that there had been at the very most four glacial episodes whereas an astronomical explanation meant that glacials would come and go throughout the Earth's history. An influential textbook[10] published in 1959 based its estimates for the age of the ice

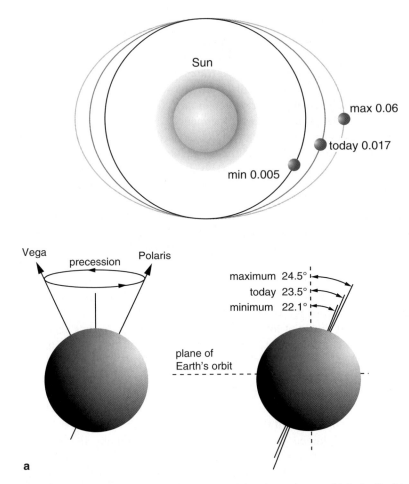

Fig. 4.2a The three Milankovitch variables: eccentricity, the extent to which the Earth's orbit departs from circularity; precession, the conical path described in space by the Earth's rotational axis; and obliquity (axial tilt)

ages on the astronomical theory, as it had come to be known, yet it stuck to a four-fold glacial scheme that had first been formulated in the Alps in 1909.

The logjam was broken in 1976, when the chemistry and fossil content of cores drilled into the sea floor were found to display evidence of 20 or more ice ages. The evidence consisted primarily of the ratio between two isotopes of oxygen, ^{18}O and ^{16}O, which reflects the extent of evaporation from the ocean (as the lighter ^{16}O escapes more freely), and thus the proportion of the world's water that is stored in ice caps and glaciers.

The sediments cored represent the latter part of the Pleistocene glacial period whose origins can be traced back perhaps 40 Myr with the birth of the Antarctic ice

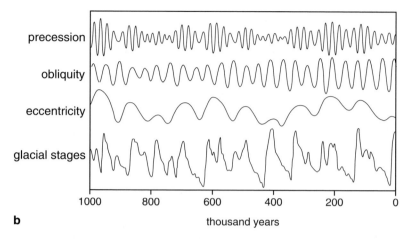

Fig. 4.2b Periodicity of the three Milankovitch variables for the Earth and glacial stages identified from oxygen isotopes in deep-sea cores

sheet but which was not clearly manifested in the northern hemisphere until 3 Myr or so ago. Four earlier, major glacials were the Huronian (2,400–2,100 Myr ago), the late Proterozoic (800–630 Myr ago), the Ordovician-Silurian (450–420 Myr ago) and the Carbo-Permian (360–260 Myr ago). Their origin undoubtedly owed much to the location as well as the topography of the landmasses. For instance what is now southern Algeria was located over the South Pole during the late Ordovician when Africa was part of the southern megacontinent of Gondwana.

The Pleistocene glacial era included some 20 or so glacial advances and retreats. The reason for the repeated oscillations is still unclear. The habit has grown of 'tuning' the data to the astronomical timescale, that is of presuming that astronomical changes drive the system. Even then only the 100,000 year cycle is clearly represented whereas it should be the least prominent of the three Milankovitch variables. Some scientists have accordingly suggested that every 100 Myr or thereabouts the Earth passes through a cloud of interstellar dust which changes the level of solar radiation incident on the Earth without the need for any other mechanism, a simple but currently untestable proposition.

There remains the vexed issue of solar variability. Major changes in solar output can hardly be ruled out when (even if we include dependable sunspot data) the observational record for our Sun spans only four centuries and that for sunlike stars a few decades at most.[11] The best place to look for a solar sequence which will reveal just how changeable the Sun can be is not in space but in the rocks, for they can be drilled and analysed billions of years after the event. For several years the oldest contender was the Elatina sequence of Western Australia, which dates from 650 Myr ago. The layers of sandstone and siltstone of which it consists appeared to reflect a repeated cycle lasting 10–14 years, thus confirming that the Sun's pulse had remained steady ever since. But these laminae are now thought to be the product of tidal cycles and therefore to embody a daily rather than an annual drumbeat.[12]

The only consolation is that, as we all know, the tides are yoked to the Moon; and a lunar calendar may again lead us to the solar signal. In other words the Elatina sequence could yet reveal some facet of the Sun's middle aged behaviour. Until then we have to be content with a few measly millennia.

Climate in history

The sunspot cycle has long been suspected of driving significant climatic cycles. In 1801 William Herschel reported that wheat prices were higher when sunspots were scarce and suggested that this was because growing conditions were less favourable – in the UK at any rate – when a weak Sun led to cooler conditions. In 1878 the economist Stanley Jevons tried to link the entire trade cycle to sunspots. In the same year a Member of the UK Parliament, Lyon Playfair, applied the Herschel model to India albeit by arguing that the sunspot cycle would produce poor harvests when it led to drought.

The last Pleistocene glacial episode ended about 16,000 years ago, its close marked by the gradual disappearance of ice sheets and glaciers which had covered much of northern Eurasia and North America and by a rise in sea level which in some regions totalled 120 m. The process of deglaciation was uneven and locally went into reverse, notably during a period, named the Younger Dryas (YD) after a distinctive plant fossil, which saw temperatures in the North Atlantic region fall by about 5°C in the space of a decade or two approximately 12,000 years ago and abruptly rise by 7°C some 1,300 years later.

Some scientists believe that the YD was caused by changes in the oceanic circulation which were triggered by the release of meltwater from the North American ice sheets flowing down the Mississippi and the St Lawrence rivers into the sea. Others have concluded that events in the oceans are inadequate to explain the observed cooling. They note that [10]Be in the ice core records indicates a sharp increase in the GCR flux at about this time and argue that, by providing nuclei around which water drops could form in the air, it led to an increase in the amount of low cloud cover and therefore to cooling of the Earth as a whole by increasing its albedo.[13]

The Younger Dryas overlaps in time a period when parts of the Sahara were occupied by freshwater lakes (Fig. 4.3) notably in the Fezzan and the Chad basin.[14] The fauna depicted in the rich rock art of the region (Fig. 4.4) attests to the prevalence of a savannah environment through much of the desert. The giraffe, rhino, and cattle disappeared abruptly.

The explanation for a green Sahara that is currently favoured is that about 10,000 years ago the Milankovitch mechanism led to a peak in insolation near latitude 20°N in the northern summer and that this led to a northward shift of the African monsoon which brought with it increased rainfall. There is evidence for two dominant wind directions at the time, one from the NE and the other from the SW, which makes sense if the monsoon penetrated deeper into the Sahara than it does now. There is evidence for an older episode of high lake levels which has not yet been dated.

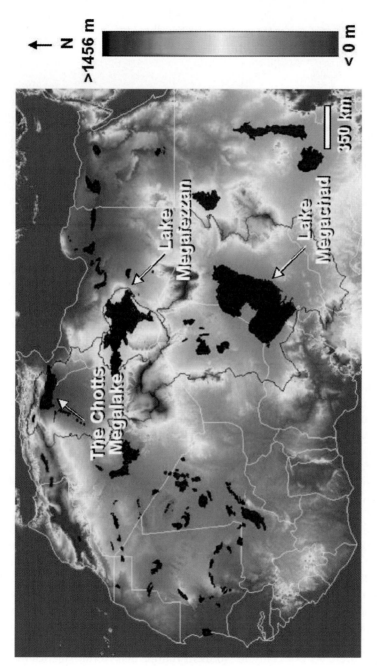

Fig. 4.3 Holocene lakes in the Sahara (After Drake & Bristow 2006, courtesy of Nick Drake)

Fig. 4.4 Libyan Sahara rock art. Giraffe and symbol probably of the Sun in Wadi Matkhandoush, Fezzan, Libya, possibly from the 8th millennium BC. Majdi Makhlof as scale

At its maximum, a little over 7,000 years ago, one of the lakes, Megachad, was over 170 m deep and occupied an area of at least 400,000 km². There are traces of an earlier lake which was twice as large. Other former North African lakes of similar age include Megafezzan and the Chotts basin (Fig. 4.3). Together they covered 10% of the Sahara, and evaporation from these large water bodies further enhanced local rainfall. The humid episode ended 5 millennia later, though more suddenly than the solar mechanism would imply, because the vegetation cover collapsed rather than gradually fading away.[15]

Even if the precise mechanism has still to be worked out there is little argument that the Sun played a part in filling and emptying the Saharan lakes. No such consensus exists over a much more recent climatic anomaly lasting several decades for which there is abundant documentary evidence and which, what is more, coincided with a well attested spell of reduced solar activity.

The term Little Ice Age (LIA) was first applied to the period roughly between AD 1550 and 1850, when mountain glaciers in both the northern and the southern hemisphere were much more extensive than now and in many instances extended 1–2 km beyond their present fronts. Over the years evidence has accumulated for many related effects especially in Northern Europe and North America including the spread of pack ice in the Atlantic, the freezing of coastal waters and rivers, harsh winters and prolonged flooding. Through much of the period between AD 1400 and

1910 temperatures in many parts of the northern hemisphere were 0.5–1.5°C lower than in the latter part of the 20th century.[16] In some areas cooling began in the 13th century and perhaps even earlier. Similar reports from Africa, China and South America show that the episode was not confined to the North Atlantic coasts.

Only one clear benefit of the anomalous climate has been documented, and we are still reaping it. Longer winters and cooler summers are suspected of reducing the growth rate of certain tree species including spruce and it has been suggested that the violin sounding boards used by Antonio Stradivari in his golden period (1700–1720) owe some of their quality to those unusual environmental conditions.[17] Otherwise the human consequences were largely dismal, with the abandonment of many settlements in Greenland, and widespread crop failure and famine in much of Europe.

The various events were broadly synchronous in many parts of the world. Yet in 2001 an international group of several hundred climatologists and other environmental scientists decreed that the term LIA had 'limited utility' for describing trends in hemispheric or global temperature changes.[18] As its utility is not inferior to any other bit of stratigraphic nomenclature, such as Pleistocene, Palaeolithic or medieval, the motive may have been to discourage solar explanations which hinge on lumping together a disparate set of minor climatic episodes. Terminological matters aside, there is no disputing that the Maunder sunspot minimum is conventionally dated to AD 1645–1715 and thus falls squarely within the Little Ice Age as originally defined. Many authors accept that the climatic connection between the two is self-evident, that the LIA is convincing evidence for the Sun's direct impact on global temperatures, and that the sunspot number is a direct measure of solar activity. They overlook the awkward fact that climate had cooled before the Maunder Minimum and had stayed cool after the MM had ended. The discrepancy is resolved if the Wolf, Spörer and Dalton minima are grouped with the Maunder to give a period poor in sunspots that stretched from AD 1280 to 1820 – in excellent agreement with the newly expanded LIA.

The missing link

There still remains the tricky matter of showing whether reduced solar output led to cooling. In many studies of ancient climate there is a failure to identify a physical link between a lack of sunspots and little ice ages, or, for that matter, sunspots and normal weather. Much excitement was generated in 1991 by reports that changes in the length of the sunspot cycle closely matched those in global surface temperature.[19] Why that should be was never satisfactorily explained, but one version of the graph comparing cycle length and temperature showed a steep increase in both during recent decades, the clear implication being that global warming was a natural, solar effect. In 2006, *EOS*, the weekly newspaper of the American Geophysical Union, carried an article entitled 'pattern of strange errors plagues solar activity and terrestrial climate data'. Its thrust was that the books had been cooked to strengthen the solar case. The language was courteous, as we see from the title of the piece, but the message fierce: *These findings do not by any means rule out the existence of important links between solar activity and terrestrial climate…The sole objective of the present analysis is to draw attention to the fact that some of the widely publicized, apparent correlations do not properly reflect the underlying physical data.*[20]

The critics argued that, whereas the cycle lengths for the earlier part of the curve had been smoothed, the four data points for the period of global warming had been left in their partially or totally raw state on the grounds that the requisite information for filtering was not available. Moreover, little had been done to put this right in an update published in 2000. Figure 4.5 shows the version favoured by the critics in which, besides some minor arithmetical corrections, the 'properly filtered' data points are shown. The correlation evaporates and the recalculated solar effect comes out at most as 25% of the total warming until 1980 falling to 15% by 1997.

The scepticism is in keeping with the general consensus. An authoritative study published in 2007 goes further by stating that, whereas 'there is considerable evidence for solar influence on the Earth's pre-industrial climate and the Sun may well have been a factor in post-industrial climate change in the first half of the last century', for the period 1975–2006 'all the trends in the Sun that could have had an influence on the Earth's climate have been in the opposite direction to that required to explain the observed rise in global mean temperatures'.[21] In other words, the solar factor at the close of the 20th century if acting alone would have led to a fall in global temperatures.

How does this square with the LIA? If we calibrate the sunspot data with satellite measurements where they overlap, the reduced solar output indicated by the

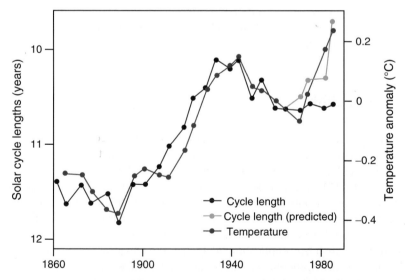

Fig. 4.5 Correlation between length of sunspot cycle and temperature anomaly since AD 1860 and alternative versions of events since global warming accelerated[20] (By permission of the American Geophysical Union)

Maunder Minimum, even if largely confined to the UV part of the spectrum, would have resulted at most in a reduction in global temperatures of 0–0.3–0.4°C, though in the eastern USA and northern Eurasia perhaps closer to 1°C. The solar contribution to the ensuing rise in temperature in irradiance, at least for the period 1900–1955, was roughly equal to that due to manmade gas forcing.[22]

That seems inadequate to explain the worst excesses of the LIA, but the effect of a modest fall in temperature could have been amplified by the cooling effect of an extensive cloud cover. Satellite measurements of the percentage of the Earth covered by low cloud today are closely correlated with the GCR flux during the same period as measured by instruments carried aloft by balloons. The inflow of GCRs, as we saw in Chapter 1, is reduced by the Earth's magnetic field and by the solar wind, and will presumably be enhanced at times of reduced solar activity. In other words the solar minimum of AD 1280–1820 could be more important for its swollen cosmic ray flux than for its direct impact on temperature.

Cosmic rays promote the nucleation of clouds, perhaps because charged particles seed water droplets. As it happens the cloud chamber, which was invented in 1899 and proved of critical importance in particle physics by revealing the tracks left by interacting particles in saturated air, was originally intended for creating clouds in the laboratory.[23]

In order specifically to test the suggestion that cosmic rays can play a crucial part in climatic change by promoting the development of low-level cloud CERN, the European Centre for Nuclear Research, has created a facility called CLOUD, a desperate acronym for Cosmics Leaving OUtdoor Droplets, which will have as its first task to investigate cloud microphysics using a beam from a particle accelerator as an artificial source of cosmic rays. At the heart of the machine is a chamber which will recreate cloud conditions throughout the atmosphere (Fig. 4.6).

In 1998, when the project was still in its early stages, its lead scientist suggested that the Sun and cosmic rays might have accounted for somewhere between a half and the whole of the increase in the Earth's temperature of the preceding 100 years. Funding was frozen in the face of protests by climate scientists that his comments would be exploited by spokesmen for the oil industry,[24] and work on the project apparently did not resume until 2007.

Shifting climates

There is a second mechanism which can amplify the solar signal. The UV part of the solar spectrum impinges mainly on the stratosphere rather than directly on the atmosphere. Absorption of solar radiation by stratospheric ozone mainly in the 200–300 nm wavelengths results in daily heating by about 8°C at elevations of 50 km in middle latitudes[25] and to peak temperatures of about −3°C near the top of the stratosphere. This heating is a major source of energy for the atmospheric circulation.

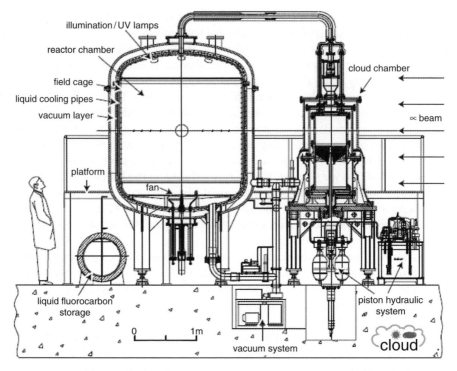

Fig. 4.6 Vertical section through the CLOUD facility at CERN showing the 0.5m cloud chamber and the 2m reactor chamber. The beam counters are not shown. (Courtesy of J. Kirkby, November 2007)

The amount of UV reaching the ground evidently depends not only on the nature of the Sun's output but also on how much is transmitted, absorbed and reflected by the atmosphere, and thus on the local altitude, latitude, time of day, and time of year. UV-B is absorbed and also reflected by clouds, pollutants and ozone in the atmosphere but, most important, by ozone in the stratosphere, 10–50 km above sea level. But the ozone that is produced when solar UV dissociates atmospheric oxygen is present in very small amounts, equivalent on average to 0.00006% by volume with a maximum at 20–30 km, and it is destroyed by reactions involving nitrogen, hydrogen and chlorine, as when the decomposition of manmade chlorofluorocarbons (CFCs) by solar UV liberates chlorine atoms. In short, heightened ozone shielding may just about offset any increase in UV at solar maximum as regards the risk of sunburn but in the present context what matters more is any associated stratospheric heating.

Net radiation during the Maunder Minimum was 0.3% lower than today and the UV (<300 nm) component about 3% lower.[26] Computer modelling suggests that heating of stratospheric ozone caused by an increase in the Sun's luminosity in the UV (175–320 nm) part of the spectrum can result in poleward displacement of the tropospheric subtropical jet streams[27] and the associated mid-latitude storm tracks.

A reduction in solar UV might therefore shift the jet streams, and their associated weather systems, towards the Equator.

The jet streams are narrow, elevated belts of high speed winds that flow roughly parallel with latitude and are typically thousands of kilometers long, a few hundred kilometers wide, and only about 2 3 km thick. They are usually found at elevations of 10–15 km. The tropical jet streams, located about 45° N and S, are weak because the temperature contrasts that occur from place to place, and between the seasons, are hereabouts relatively slight. More prominent are the polar night jets, named after the 6 month polar 'nights' which afflict the hemispheres alternately and which result in a steep temperature gradient between the area poleward of roughly 60° and the part of the stratosphere subject to ozone heating.

The contrast is especially pronounced during southern hemisphere winter because the Antarctic is colder than the Arctic and there are no massive mountain chains or land/sea contrasts to disrupt the circulation pattern. The northern jet stream is more prone to departure from a clear W-E path: whereas the southern hemisphere winter wind field is very zonal, that is to say aligned mainly E-W, the northern hemisphere winter wind field has a noticeably undulating (or meridional) geometry. As you move poleward in the southern hemisphere winter, therefore, the gradient in the wind field is much stronger than in the northern hemisphere winter.

The jet streams influence the location and severity of weather patterns beneath them. When flow is markedly meridional, it gives rise to pressure ridges and troughs, and thus high-pressure and low-pressure systems. The mid-latitude jet stream determines the strength and track of the depressions that dominate the weather of western Europe. A glance at summary storm maps for winter in the middle and at the end of an 11-year sunspot cycle (December–February 1991–2 and 1995–6) shows how this effect is translated into contrasting storm track patterns (Fig. 4.7).

Added complications may be introduced by the El Niño effect, named after the Child (i.e. Christ) because it tends to occur around Christmas. It sees a warming of waters in the eastern Pacific Ocean near the Equator. Normally, the trade winds blow westward, carrying warm surface waters to Indonesia and Australia and allowing cooler waters to well up along the South American coast. Every 3–7 years the trade winds are weakened or even reversed, so that warmer waters move toward the coast of South America and raise water temperatures there. One consequence is that the polar jet stream over North and South America splits in the North Pacific, diverting storms toward the Yukon and Northwest Territories, while leaving most of Southern Canada with a milder and drier-than-normal winter. The El Niño may even influence the weather in Europe: in 1997 and 1998, for instance, it brought record high winter temperatures.

During a La Niña the converse occurs. In the North Pacific, the polar jet stream frequently takes a path northward toward the Bering Straight and then to the southeast, toward the Pacific Northwest. In the meantime, a vigorous subtropical jet moves northeastward across the Pacific, commonly passing over the vicinity of the Hawaiian islands.

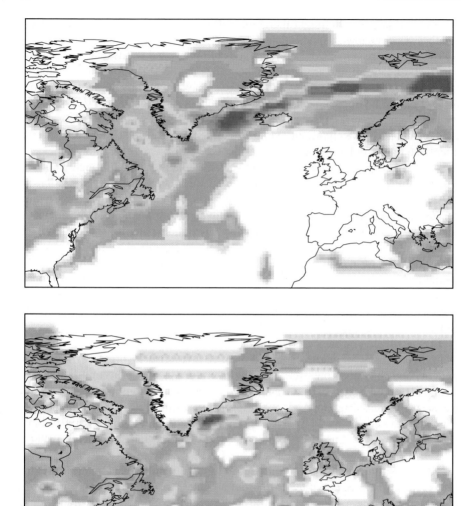

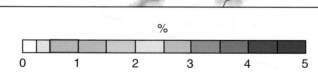

Fig. 4.7 Storm track frequency in the North Atlantic for December–February 1991–2 and 1995–6 to illustrate conditions at maximum and minimum of the solar cycle (Plots courtesy of Mark Chandler and Jeff Jonas, Columbia University, Goddard Institute for Space Studies)

Shifts in the position of the main global weather systems have long been documented by mariners and cartographers. In the 16th and 17th centuries the westerlies appear to have temporarily shifted south over the eastern Pacific.[28] During 1921–1976 the cyclone tracks in the North Atlantic north of 50° moved south an average of 2.5° and a maximum of 6° during sunspot maxima.[29] A similar equatorward shift has been detected in the southern hemisphere.[30]

In the Mediterranean basin, destructive soil erosion is currently driven by heavy summer rains acting on poorly vegetated soils. In the period between about AD 500 and 1900 the rivers of the region experienced a phase of silting which, besides burying entire cities such as Olympia in Greece, created large areas of malarial, waterlogged terrain. In places the depth of silt exceeded 10 m. Many Classical sites besides Olympia were buried. In the Po valley several cities were devastated by floods and filled with mud.

The field evidence confirms that the flow pattern in the rivers has since changed. The geometry, texture, geochemistry and palaeontology of the historical deposits all point to a period when channel flow was less ephemeral and irregular than it is today. The evidence includes a temporary steepening of the channel long profiles, presumably to counteract weaker peak discharges and thus maintain the balance between water energy and sediment load; the prevalence of well bedded, coarse silt and sand reflecting sustained flow where older deposits point to shortlived floods; and a structure that is inconsistent with the effects of manmade erosion.

The silting phase appears to have spread gradually southward across southern Europe and SW Asia. At any one location it lasted a few hundred years. But the downcutting that brought it to an end was swift. The broadly synchronous timing of aggradation throughout the Old World, with its complex cultural history, is also more consistent with a climatic factor than with a change in agricultural practice or widespread abuse and neglect of the land. Yet many scholars continue to ascribe the erosion that supplied the valley silts to human activity even where the evidence for it is lacking.

Dating of the deposits hints at the gradual spread of the silting phase southward across southern Europe and SW Asia. The silting might last several hundred years. But the downcutting that brought it to an end was swift. The simplest explanation for the silting is a shift in the timing and in the relative importance of erosive rains. This is consistent with what occurred during 1961–1998, when the tracks followed by extratropical storms at solar maximum were concentrated in the extreme north of the Atlantic and at solar minimum affected large parts of the Mediterranean and the Near East.[31] The range of available ^{14}C ages for the deposits matches the growing evidence for a succession of anomalous conditions going back at least to the 14th century AD[32] rather than (as formerly thought) a single Little Ice Age during the 17th–18th centuries.

The solar model gains further support from evidence for a very similar unit in southern North America The first reports[33] described a depositional unit which was represented in Nevada, Wyoming, Colorado, Arizona, and New Mexico, and for which the ^{14}C ages then available indicated deposition between ~1,600 and 200 years ago. Similar trends are reported from the southern Colorado Plateau, coastal California, Texas, Nebraska, Montana and Mexico. The details vary from place to place and there are interruptions, but, granted that there are likely to be variations in the timing of alluvial deposition in different valleys even in the same river basin because both filling and cutting move progressively up- and downstream,[34] the general picture is strikingly uniform.[35]

Throughout the southwestern USA the white settlers had encountered extensive, well watered valley floors which were cultivated by Indian farmers using low dams and dikes to spread the waters across their fields. The colonists' wagons were able to cross the valleys with little difficulty. Starting in the late 19th century many of these flats were rapidly trenched, to become impassable by wheeled traffic and almost impossible to irrigate without the aid of dams or pumps.

The reason for the widespread channel cutting has long been disputed. As in the Old World, the two main suspects are climate and human activity. The answer is probably a bit of both: clearing the vegetation on the valley floors, overgrazing and trampling by herds encouraged erosion, and a change towards short sharp storms at the expense of gentle rains did the rest. The impact of changing rainfall patterns on channel behaviour is especially clear in central Mexico, where the proportion of the annual rain that falls as erosive thunderstorms shifts progressively north and south with the subtropical high-pressure systems.[36]

In other words, slight changes in the UV component of solar radiation may account for shifts in the position of the climatic belts which in the Middle Ages resulted in dramatic slope erosion, soil loss and valley silting (Fig. 4.8). As a result, cities were filled with glutinous mud – and much, of the richest agricultural land in both the Old and the New Worlds was laid down. As with the Saharan lakes, the Sun appears to have played a discreet but crucial part in the story.

Fig. 4.8 Fills in Louros valley, Epirus (Greece). The cultivated, flat-topped deposit in the middle distance was deposited in medieval times in response to a shift in rainfall patterns

Chapter 5
Sun and Life

Water and sunlight, it is often claimed, are the prerequisites for life on Earth.[1] Liquid water allows and in some instances encourages chemicals to combine into more complex forms, and it was crucial to the emergence of life by favouring the diffusion and exchange of organic molecules.[2] And of course many plants and simple organisms obtain their energy from the Sun by photosynthesis.

In 1977 the scientific crew aboard the 8 m-long submersible Alvin was investigating the sea floor off the Galápagos Islands at depths of 2,500 m. To their surprise they found that the local black smokers, the informal name for geothermal vents which give off mineral-rich water at temperatures of over 400°C, supported a rich population of giant clams, crabs and tubeworms (Fig. 5.1) well away from sunlight or the atmosphere. Hydrothermal vents have since been identified on the southern Mid-Atlantic Ridge, on the Carlsberg Ridge in the Pacific, and off Papua New Guinea. Fossil hydrothermal locations are also known, the oldest so far being a deposit in North China which is 1.43 Gyr old.

Early days

The significance of the strange submarine life forms was not immediately grasped. It took a graduate student, Colleen Cavanaugh, to explain the mechanism during a lecture she was attending on the fauna. She interrupted to suggest that the clams and tubeworms depended for their food on specialized bacteria within their tissues which get their energy from processing hydrogen sulphide (H_2S) from the vents in the same way as plants use sunlight. The lecturer told her 'No, no, sit down, kid. Hydrogen sulphide is a potent toxin'.[3] He soon had to eat his words.

In parentheses note that neither side appears to have done its homework. A classic biology textbook first published in 1924 states 'Certain bacteria whose metabolism is based on iron, sulphur or selenium derive their energy from other sources. They are thus independent of sunlight – a fact of the greatest significance in connection with the problem of the origin of terrestrial life as we know it today.'[4] It was not long before other sunless environments began to yield thriving communities of extremophiles which tolerate physical and chemical conditions that had not long

C. Vita-Finzi, *The Sun: A User's Manual*,
doi: 10.1007/978-1-4020-6881-2_5, © Springer Science+Business Media B.V. 2008

Fig. 5.1 The tube worm *Riftia pachyptila*, which uses chemosynthesis instead of light for energy, lives near hydrothermal vents on the floor of the Pacific Ocean (Courtesy of NOAA)

ago been considered lethal. Mineral exploration revealed microorganisms at depths of 3 km within the Earth's crust. Some of them live off trapped or escaping gases such as methane; some manufacture their own food from rock minerals; some, like the tubeworms, are dependent on other organisms. These endoliths live under high pressures up to more than 1,000 atmospheres, the barophiles through necessity, the barotolerant with a shrug. The Archaea, single-celled microorganisms which before 1977 were lumped together with bacteria but are now considered distinct, include *Methanogenium frigidum*, which lives in Ace Lake, Antarctica, in waters that are permanently at 1–2°C.

Besides demoting the Sun as unique energy source these discoveries have reduced the problems created by the faint young Sun. Extremophiles have also widened the range of environments that are considered in the search for the origins of life on Earth and on other bodies in the Solar System. Nevertheless most known extremophiles are active within the small range of −2° to +50°C even with the help of such stratagems as the use of glycerol and other kinds of antifreeze, whereas temperatures in the bulk of the universe lie outside the 0–100°C range.[5]

The plot was bound to thicken. In 1989 a small eyeless shrimp from one of the Atlantic vent fields was found to have light-sensitive pigments on its back. And green sulphur bacteria which are normally photosynthetic have been found at depths of 100 m in the Black Sea, suggesting that they exploit near-IR radiation, which is given off by anything above absolute zero (the zero on the Kelvin scale). But the hydrothermal vents also emit very faint visible light, possibly produced from chemical reactions or crystal breakage, and the bacteria possess chlorosomes, devices that scoop up the few photons to be had.[6] In other words, light, though not sunlight and not necessarily visible to us, still appears desirable for submarine life, and photosynthesis appears a sensible process even when the effort seems disproportionate. It remains to be seen whether the Earth's crust will also be found to be luminous.

Work done in 2007 has added a sinister twist to the bacterial links with light. It appears that some bacteria, including the species of *Brucella* responsible for brucellosis or Malta fever, become virulent when exposed to blue light. The sensors that provide the link are closely related to the light receptor molecules that prompt a plant to grow towards a light source[7] and that are acronymised as LOV for light, oxygen or voltage domains.

Whatever the real significance of light, water is still widely considered of crucial importance to life, and the search for life elsewhere in the universe focuses on the habitable zone around stars where water can remain liquid. For hotter stars the minimum radius will be further out than for our solar system, and for cooler, red stars, much further in. Whence the excitement that greeted the discovery in 2007 of the unmistakable signal of water on planet HD 189733B.

But even water is losing its privileged position. In 2004 when the Cassini probe was orbiting Saturn's moon Titan, the low temperatures there seemed to rule out life. But dissenting voices argued that the fundamental requirements are an energy source, which could be radioactive decay just as well as the Sun, and a range of temperatures which permits chemical bonding. On Titan, with its methane lakes, hydrocarbons could substitute for water as solvents in managing complex organic reactions.[8]

In a famous letter to his botanist friend Joseph Hooker, Charles Darwin[9] mused over the origins of life:

> If ....we could conceive in some warm little pond, with all sorts of ammonia and phosphoric salts, light, heat, electricity, etc., present that a protein compound was chemically formed, ready to undergo still more complex changes...

Half a century later nutritious soup was still on the menu. The critical step was taken in 1953 by Stanley L. Miller[10] when he exposed a mixture of ammonia, methane, hydrogen and water to electrical discharges to simulate lightning. In the organic compounds that accumulated in the space of a week Miller identified 13 of the 20 naturally occurring amino acids, the building blocks of proteins.

The Miller experiment remains controversial, as there are many slips between inert amino acids and replicating life; and the reducing conditions postulated by Miller are not universally accepted. But the fact remains that here was a mechanism which generated organic material in a rather humdrum fashion. In the words of the astronomer Carl Sagan, the experiment 'is now recognized as the single most significant step in convincing many scientists that life is likely to be abundant in the cosmos'.[11]

Miller used an electric spark rather than UV because UV rapidly degrades amino acids. If the young stars observed from the Chandra orbiting X-ray observatory[12] are any guide, the Sun for its first 50 Myr was highly unstable, and for the next 500

Myr it emitted X-rays (0.01–10 nm) and other damaging radiation at levels estimated at 50 times higher than today's. In other words the faint young Sun was
throwing tantrums just when the building blocks of life were first being
assembled.

Just how strong and harmful was the level of UV radiation on the young Earth?
It is not an easy question because the answer depends on the Sun's luminosity, the
composition of the Earth's atmosphere, and, once some shielding ozone was in
place, the extent to which the ozone was damaged by volcanic eruptions, meteorite
impacts and supernova explosions.[13] At any event, modelling studies[14] suggest that
radiation at <200 nm (i.e. X-rays, gamma rays and part of the UV-C range) was
blocked by what little atmosphere there was before 2.2-2.0 Gyr, when oxygen and
therefore ozone levels began to build up significantly, whereas the flux of UV radiation as a whole even with a weaker Sun was perhaps as much as 10^{30} times higher
than today.

Now UV-C was shown earlier to combine the roles of midwife and nanny insofar
as it promotes ozone formation in the atmosphere but it is also strongly absorbed
by stratospheric ozone. It has also acted both as mutagen, that is as a source of
mutations, and as a positive selective agent.[15] Selective pressures were strong
because the UV wavelengths in question (200–300 nm) are close to those most
readily absorbed by the nucleic acids in DNA. This is an effect that we encounter
repeatedly in our solar universe: dosages and wavelengths lurch from being beneficial to lethal within dangerously small margins.

There were two early key survival strategies: protecting the genetic material by
bases or nucleotides which had no function for replication or for protein synthesis;
and lurking in rock crevices or in ocean waters at depths of over 50 m, at which the
radiation would be sufficiently attenuated.[16]

Every cell must face a constant onslaught of DNA lesions. Many of them arise
from the workings of the cell itself, rather than UV-induced damage, so that it is
fair to speak of a DNA half-life; and in order to avoid harmful mutations, a
destabilised DNA sequence, or cell death, an estimated 10,000 lesions (amid the
3 billion base pairs in the genome) are dealt with each day.[17] Even so solar UV
remains a major potential source of DNA damage, alongside ionizing radiation
such as X-rays and gamma rays, and cosmic rays.

The lesions are countered by inbuilt repair mechanisms, as we will see in a
moment with the bacterium *Deinococcus radiodurans*; by sex, which may allow an
undamaged copy of the DNA to be recovered; and by letting natural selection run
its course so that those individuals that cannot cope are eliminated.[18]

Moreover there is a school of thought which sees UV radiation, rather than
lightning, as the crucial energy source for the synthesis of large molecules from the
primeval broth. Indeed, experiments carried out in the 1960s found that irradiation
by UV promoted the formation of ATP, an energy-carrying molecule which is found
in the cells of all living organisms.[19]

The need to strike a balance between selective protection from UV and access to
other wavelengths of solar energy governed the lifestyle of the oldest substantial
fossils, the banded, mushroom-shaped structures known as stromatolites. The earliest

Fig. 5.2 Stromatolites, colonial structures created by cyanobacteria, in Shark Bay, Australia (Courtesy of Warwick Hillier)

specimens date back over 2.7 Gyr. Theirs was a winning formula, witness the modern examples in Australia and elsewhere (Fig. 5.2). The structures are built up by micro-oganisms, especially cyanobacteria, which convert CO_2 and water into nutrients with O_2 as a waste product. Lower in the mounds are to be found anaerobic bacteria to which oxygen is a poison.[20] Cyanobacteria are sensitive to UV-B, and some have evolved screening pigments, and it may be that the mucus that traps calcium carbonate and that leads to the layering of stromatolites has a protective role.

When shelter from harmful UV was sought in the ocean the longer wavelengths of visible light were filtered out and the only wavelengths available for the earliest photosynthetic pigments were presumably the shorter ones. This, as well as its efficiency, may explain the dominance in photosynthesis of blue light with wavelengths of about 430 nm.[21]

Nevertheless chlorophyll, the most widespread light-absorbing pigment in plants, has a second peak in the red part of the spectrum at about 660 nm. It thus makes little use of green light (~550 nm), which is the most energetic part of the visible spectrum and is indeed the part to which the human eye is most sensitive. A possible explanation is that early photosynthesis exploited the red (as well as the IR) light emitted by submarine hydrothermal activity.

UV radiation has even been brought in to explain the chirality of living matter. In 1849 the bacteriologist Louis Pasteur discovered this effect when he noted that crystals of tartaric acid came in two forms which were mirror images of each other. When he had sorted them into two groups and dissolved them he found that one of the solutions rotated polarized light (that is to say light in which the waves are transmitted in only one plane) clockwise and the other solution anticlockwise. Protein molecules in living organisms are composed of 20 amino acids. In proteins they exist in the left-handed form (laevorotatory), with the exception of glycine, which is symmetrical; and in sugars the molecules are dextrorotatory. Macromolecules like DNA follow the protein scheme. As replication and many functions depend on the shape of the molecule such consistency is understandable.

Why does the helix of protein amino acids wind to the left? The manufacturer of a series of skin care products claims that optically correct treatments contain the keys that unlock the skin's potential, but the evolutionary advantages of chirality are not known beyond the need to conform to the prevailing handedness. Three rival explanations are on offer: chance; selective destruction by some natural disaster of an ancestral population with right-handed amino acids,[22] and the effect of polarized UV light in the nebula out of which the Sun and planets eventually condensed.[23] The last explanation has been offered for the very existence of substances from animal and plant sources which are 'optically active', the effect investigated by Pasteur.[24]

Getting there

Any discussion of exposure to UV has to consider where the organic molecules originally came from in the first place. All the constituents of life were generated in the interiors of stars, the sole exception being hydrogen, hence the hollywoodean claim that we are all stardust. The generation of organic chemicals, once thought implausible in space, can now be explained at least in part by the action of UV. In one set of experiments the irradiation of interstellar dust grains by UV produced 16 amino acids; the amino acids survived the harsh conditions they encountered in Earth orbit provided they were embedded in clay or meteorite material.[25]

Findings of this sort have fuelled the revival of panspermia, yet another notion touted by Anaxagoras over 2,000 years ago, the belief that (in its modern guise) interstellar dust and comets have seeded life on Earth. The idea has attracted the support of the astronomer Fred Hoyle (whom we last met at Stonehenge)n and the co-discoverer of the structure of DNA Francis Crick. As is often sourly observed, however, acceptance of panspermia only defers the question of origins in time and shrugs it off onto another planet or solar system, but it has to be considered here for the bearing of solar history on its feasibility and more generally for opening our minds to the likelihood that life is common in the Universe.

The journey between two bodies in our solar system might take a few months and between two solar systems several million years. The conditions that the seeds or spores would encounter include a high vacuum, exposure to GCRs and SCRs, temperature extremes, dynamical stress, extreme dryness, chemical attack and, of course, solar radiation at various wavelengths.

Some understanding of what can be endured comes from experimental studies on spacecraft or with high velocity guns. Bacteria of the genus *Rhodococcus* have been shown to survive impacts at over 5 km/s, similar to the escape velocity for Mars.[26] Of a set of samples of salt-loving microbes (halophiles) which spent

2 weeks attached to a satellite, 10–75% survived the journey.[27] Exposure to space of spores of *Bacillus subtilis* from the NASA Long Duration Exposure facility for nearly 6 years showed that solar UV radiation was the most harmful hazard they encountered and that it reduced survival by 10,000 times or more though without wholly wiping out the sample population.[28] What ensured survival is not certain but there is at least one microbe which can undo severe damage to its genome: after desiccation and exposure to the ionising radiation the DNA of *Deinococcus radiodurans* was shattered into hundreds of short fragments but within a few hours these were accurately reassembled into a functioning sequence.[29] Of course the chances of survival for spores and other microscopic life forms are improved if, rather than being stuck to the surface of a spacecraft, they are conveyed inside meteorites or attached to dust particles which are propelled by radiation pressure from the Sun.

Ice and waves

By the middle of the Proterozoic, the name for the period from 2.5 Gyr to 540 Myr ago, before complex life became widespread on Earth, there is chemical (isotopic) evidence for oxygen, and, as you might expect, for the photosynthesis that created the oxygen. Once a substantial oxygen-rich atmosphere was in place and life was securely launched on land and sea the solar factor was felt in evolution primarily through climatic fluctuations and changes in the distribution and character of land and sea. UV, though still the most potentially lethal part of the EMR spectrum at the Earth's surface, was kept at bay by the ozone in the stratosphere; and X-rays and other ionizing forms of radiation, as in earlier times, were blocked by the atmosphere.

Several mass extinctions – the relatively rapid elimination of many diverse organisms on a global scale– have been recorded in the fossil record, two between 4.6 Ga and 540 Myr ago and five in the Phanerozoic, 'the eon of multicellular life', roughly 440, 360, 245, 208 and 65 Myr ago.

Some theories cite the motion of the Sun through the Milky Way as the key to these events. Using the high flown language of astrophysics one group of investigators put it this way: the galaxy is a thin disc, like a Frisbee, but the galaxy's motion is like a pie in the face.[30] Its passage through the north boundary of the galaxy every 62 Myr or thereabouts exposes the Earth to a large dose of high-energy radiation which increases the risk of mutations and also creates a blanket of clouds which cool the Earth possibly to the extent of precipitating prolonged ice ages.[31]

Many of the numerous alternative explanations for extinctions or major faunal changes depend on climatic change or sea-level fluctuations or both and thus embroil the Sun. The current favourite for the Cretaceous-Tertiary event, which saw the elimination of all land creatures weighing over 25 kg including all non-avian dinosaurs and pterosaurs, is an asteroid impact. A likely consequence of the blast was a large cloud of dust which would have blotted out the Sun and thus shut down photosynthesis for at least 10 years, with lethal effects on much plant life and its dependants.

Even away from the drama of mass extinctions the history of life is marked by the displacement and replacement of faunas and floras which were prompted by changes in climate, topography and sea level. No doubt many of the grander events were driven by forces from inside the Earth, especially plate tectonics, which constantly redrew the three dimensional map of the continents. But, as is often said, no water, no plate tectonics: liquid water governs the mechanical behaviour of the rocks involved in plate movements, and the atmospheric circulation is responsible for the progress of climatic history.

The response of mobile creatures to advancing glaciers or encroaching seas is generally to move out of the way, and site excavation accordingly often reveals a sequence of occupations and abandonments which may eventually, be linked into a broad sequence of physical and cultural changes. Whether *Homo sapiens* appears as the glorious culmination of an inevitable stairway of progress is largely a matter of taste; yet the environmental input to hominid evolution often remains just as controversial.

Before seaworthy vessels were devised, many habitable areas became accessible through changes in the level of the sea that created land bridges and marine corridors. The significance of seaways as barriers is in itself contentious. To begin with, there is a common reluctance to accept that early man was capable of crossing substantial bodies of water using rudimentary boats or rafts. Artefacts of Middle Palaeolithic type in Cyprus, for example, are dismissed as recent on the assumption that the island remained inaccessible until the Neolithic. Yet the Turkish coast was less than 70 km away when sea level was 120 m lower than now and we know that. Australia was colonised by humans at a time when distances of 100 km of open water had to be negotiated. Similarly, the Red Sea is viewed as a persistent barrier to movement out of Africa (Fig. 5.3) when its southern extremity – the Bab el Mandab – was probably dry land in glacial times, and even now it is now divided by an island into two stretches of which the longer measures no more than 25 km.

Take the potential effect of sea level changes on routes and resources. Before seaworthy boats were created, many habitable areas became accessible through changes in the level of the sea which created land bridges and marine corridors. The significance of seaways as barriers is in itself contentious. To begin with, there is a common reluctance to accept that early man was capable of crossing substantial bodies of water using rudimentary boats or rafts. Artefacts of Middle Palaeolithic type in Cyprus, for example, are dismissed as recent on the assumption that the island remained inaccessible until the Neolithic even though the Turkish coast was less than 70 km away when sea level was 120 m lower than now and we know that Australia was colonised by humans at a time when distances of 100 km of open water had to be negotiated. Similarly, the Red Sea is viewed as a long-standing barrier to movement out of Africa (Fig. 5.3), yet its southern entrance – the Bab el Mandab – was probably dry land in glacial times and it is now divided by an island into two stretches of which the widest measures no more than 25 km.

Perhaps the most contentious waterway is the Bering gap, which has long been seen as a major obstacle to human migration from Asia into the Americas: even though the Straits, which now measure less than 90 km, were also dry land during the last glacial maximum, immigration was supposedly delayed until 12,000 years ago.

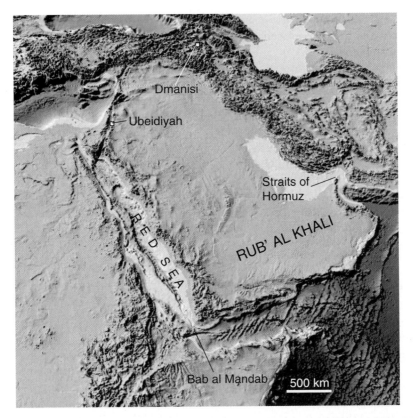

Fig. 5.3 The Red Sea. A fall in sea level would make it easier to cross at almost any point. When sea level was like the present the two easiest are the Bab al Mandab to the south , which leads to the Straits of Hormuz and thence Iran, and the Nile delta, which leads to the Jordan Valley and the Caucasus

The case for a late crossing rests mainly on the ^{14}C dates obtained for the earliest American kill sites, many of them associated with artefacts of the Clovis culture and in the region of 12,250–13,110 years ago.[32] Yet there is no real reason why movement should have been delayed until the late Pleistocene. It is often easier to walk on ice than water, witness the fact that the frozen sea was crossed on skis in 1998 and on foot in 2006.

Indeed, a much tougher barrier in glacial times was presented by the Cordilleran part of the North American ice sheet, and its disappearance would have encouraged migration from Asia whether or not sea level had changed significantly. Moreover this may well have happened on several previous occasions, and there are hints of early human occupation in the location and character of crude quartzite choppers and points, for example in Wyoming.[33] The site of Monte Verde, in southern Chile, dates back at least to 14,000–14,500 years ago, that is 1000 years before Clovis.[34] The blitzkrieg model of animal extinctions, which sees humans sweeping down to Tierra del Fuego in less than a millennium, en route slaughtering to extinction some

30 genera of large animals including mammoth, mastodon and native camels,[35] would just have to be replaced by a slower, more prosaic though no less dispiriting history of overkill.

Isolation, which seems to provide a powerful device for evolutionary innovation is as hard to prove as accessibility. One of its consequence where resources are limited has long been held to be dwarfism. Good examples of this effect on evolution come from the Mediterranean, where sea-level fluctuations during the last million years repeatedly cut off any source of genetic replenishment from various islands. Dwarf elephants have been found on Malta, Cyprus, Sicily, Sardinia and Crete; the dwarf elephant of Cyprus weighed about 200 kg. Crete and Cyprus also hosted a dwarf hippopotamus. Similarly, dwarf mammoths evolved on Wrangel island in Siberia, where they survived until the 2nd millennium BC, and on the Catalina islands of California. The islands were colonised afresh whenever sea level fell sufficiently low, so that dwarf species may have evolved repeatedly.

Human origins

Dwarf elephants have also been found in some Indonesian islands including Flores. Here in 2003 anthropologists found the skeletal remains of hominid remains ranging in age from 94,000 to 13,000 years which indicated a small body and a small brain, with artefacts of appropriate size. The remains were initially dismissed as those of a microcephalic individual but the brain of Flores man does not fit that suggestion and the consensus now recognises a new species: *Homo floresiensis*.[36] The suggestion is that *H. floresiensis* reached the island about 850,000 years ago and gradually shrank to a height of about 1 m. Whereas the local elephants also became dwarfed, lizards and rodents flourished.

These different responses support the dull but sensible view that dwarfism is not inevitable and that size depends on the nature of each island, the biology of the species in question[37] and, rather like the handedness of the original organic molecule, an element of pure chance.

The issue of adaptation is even more frustrating when the adaptative strategy (or physical response) cannot be identified with certainty from the fossil evidence because it is skin deep metaphorically if a matter of behavioural cunning and literally if an anatomical effect that is not commemorated in the shape or size of the skeleton. But there is encouraging progress on this front. Genetic studies of a 43,000 year old Neanderthal individual from El Sidrón in Spain and another, 50,000 years old, from the Monti Lentini in Italy suggests that they had red hair and pale skins.[38] The finding could reflect the tendency for populations away from the tropics to have reduced melanin levels in their skin. Some commentators have noted that the remains of woolly rhinos and mammoths in the permafrost also have reddish hair.

A link between pigmentation in humans and latitude has long been accepted but its physical basis remained uncertain. In a textbook published in 1977, for instance, the geographical relation between UV light and skin colour is explained by the protective effect of dark pigmentation, specifically pigmentation of the corneum, against skin cancer and disabling sunburn. Conversely, deep penetration of UV into the skin appeared to account for the prevalence of pale skins at high latitudes except where, as in some Arctic areas, clear skies and strong reflection from snow and ice expose the individual to high levels of UV.

A later study which was based on the global distribution of UV levels derived from satellite measurements found a strong correlation between skin reflectance and UV radiation levels, with a maximum at 545 nm. This is near the absorption maximum for oxyhaemoglobin, which suggests that melanin pigmentation in humans has as its main function to control the effect of UV radiation on blood vessels in the skin.[39] According to this study, early in the evolution of the genus *Homo* a dark skin protected humans against UV damage of sweat glands. It also inhibited photolysis, or destruction by light, of the water soluble B vitamin folate which is essential for normal development of the embryo and for spermatogenesis.

Hence selection would have acted powerfully on migrants heading north from the tropics, and although movement along coasts and rivers initially meant that fish could provide supplementary vitamin D, pale skins came to be favoured. It is claimed that this can be seen where children with British ancestry are often blond even though their parents have darker hair and skin. But not all the migration was into sunless latitudes. Asia, Australasia and in due course the Americas were colonized, and they presented their own solar challenges. It may be that climatic (and related sea-level) changes were more important for the availability of an adequate diet than for the limits they imposed on mobility. For instance a fall in sea level might expose shellfish beds whereas a rise may make it easier to penetrate a forested or steep landscape by drowning the lower reaches of watercourses.

The Neanderthalers were present in Europe and parts of Asia for at least 200,000 years and so experienced two or more glacial cycles, but, as D. McMiken has noted, red hair and pale skins however advantageous at times of reduced UV insolation, were not disadvantageous at other times and may well have been retained by sexual selection. If you don't use it, in this case, you don't necessarily lose it.

Another adaptive item of behaviour is bipedalism, the ability to walk on two legs. It has been explained in various ways, including the manipulation and carrying of tools and weapons and the ability to spot predators in long grass. A third suggestion is that its function was primarily thermoregulatory. As forests shrank, hominids experienced greater heat stress from loss of shade. *Australopithecus* countered it by adopting an erect posture and shedding body hair.[40]

Visible light, which dominates the Sun's output as seen from Earth, encouraged the development of visual systems from light-sensitive areas of the skin to the wide range of eyes known to zoology. Most mammalian eyes have cone receptors which are sensitive to short and to medium-to-long wavelengths (dichromacy). Humans have a third cone type (trichromacy) which makes it possible to discriminate between red and green; our lenses are constructed to filter out UV wavelengths, so that (according to Arthur C. Clarke) individuals whose lenses have been replaced with artificial material following a cataract operation should be able to see in UV light where the rest of us are in total darkness[41]. Some marsupials, however, enjoy a version of trichromacy which includes sensitivity to UV wavelengths;[42] IR detection is used by rattlesnakes, boas and a number of insect species; and bees are sensitive for their navigation both to UV light and to polarised visible light.

Other, behavioural kinds of adaptation to a prevailing solar regime include the circadian rhythms (from the Latin for approximately and day) that enable many organisms to anticipate periodic environmental changes or to perform regular tasks even when the external clues are absent. One suggestion is that the earliest circadian rhythms protected cells from harmful solar UV radiation by keeping to a minimum outside activity at times of peak exposure to sunlight.

This adaptation was sabotaged by changes in the length of day. As already noted, a gradual slowing down of the Earth's rotation has resulted from the frictional effect of the daily tides as they rub on the seafloor. The interplay between the solar wind and the Earth's magnetosphere would contribute to the deceleration, which is confirmed by data on the timing of solar and lunar eclipses recorded in Babylonian and Arabian astronomical tablets. The day increased in length by 2.4 ms every century between 700 BC and today, equivalent to a 15 h day 4 Gyr ago, so that early exposure to UV was shorter than now but so was the time available for DNA repair and other kinds of recovery. It follows that some circadian patterns acquired early in the history of life no longer fit the environment or other context to which they were originally tuned. Those patterns that are reset by exposure to light may fare better.

The issue of adaptation confronts historians as well as evolutionary biologists. In the early decades of the 20th century several thinkers pondered over the rise and fall of advanced societies. The American geographer Ellsworth Huntington wrote a number of books, of which the most influential was *Mainsprings of Civilization* (1945), which argued that changes in the Sun influenced human health and intellectual activity as well as climate and thus controlled the fate of the Roman Empire.[43] The Sun did so directly through the productivity of the land and its influence on the Roman mind, and indirectly by triggering droughts in the Asian hinterland and thus precipitating devastating nomadic invasions. By the 1950s and 1960s this kind of thinking was roundly condemned as crudely deterministic and by implication racist, as the notion that civilisation is favoured by cooler climates appeared to brand the humid tropics and its inhabitants as inherently inferior.

Recent decades have seen a return to environmental determinism, although it is expressed in more acceptable language.[44] As in Huntington's work, solar variability is seen as the key factor in promoting economic and social events through the agency of climatic change. Thus droughts which lasted several decades or even centuries are blamed for the collapse of the Classic Maya Empire in about AD 750–900, the Akkadian of Mesopotamia about 2200 BC, the Tiwanaku of the Bolivian Altiplano about AD 1100, and the Moche of coastal Peru about AD 600, on the grounds that they led to social collapse and the abandonment of urban sites. As the nineteenth century geologist Charles Lyell[45] once said about catastrophism, the belief that Earth history was interrupted (and explained) by cataclysmic events, *Never was there a dogma more calculated to ... blunt the keen edge of curiosity.*

Chapter 6
Sun and Health

Sunshine helps to beat breast cancer declared the front-page headline in *The Times* of London on 4 August 2007. According to the authors of a study at Creighton University, a Jesuit University in Omaha, Nebraska, the risk of contracting skin cancer from sunbathing is drummed into women year after year, yet post-menopausal women who stay out of the sun increase their risk of developing breast cancer because the levels of vitamin D in their bloodstream may drop too low. Some of the deficit can be made up by taking vitamin D and calcium supplements but doing so reduces the risk at best by about a half. Calcium had been included in the tests because the study was initially aimed at osteoporosis, and calcium turned out to have some protective effect against cancer but it was not as strong as that from the vitamin.

Five days later the same newspaper reported that the preceding decade had seen a sharp increase in the cancers that are linked to lifestyle. According to Cancer Research UK, the body responsible for the statistics, between 1995 and 2004 the biggest rise (43%) was in malignant melanoma, which can spread (or metastasize) to other parts of the body. The consensus is that malignant melanoma is caused mainly by exposure to UV radiation in sunlight.

There is no simple rule for determining the right exposure to the Sun. It depends on when, where, how and who. But a grasp of the underlying principles undoubtedly makes it easier – and much more interesting – to devise a prudent strategy for life out of doors or in space.

Wavelengths

The wavelengths that are critical to human health are largely in the UV range. They account for about 8% of the solar radiation that impinges on the outer atmosphere but, as we saw, this value varies during the 11-year cycle by 5–10% at 200 nm and at the shortest UV wavelengths a hundredfold or more. If the large, century-scale fluctuations in ^{14}C reported earlier in this book represent processes in the Sun rather than in the Earth's magnetic field the UV component could vary even more substantially. In photobiology, that is biology with light as its primary concern, irradiance

C. Vita-Finzi, *The Sun: A User's Manual*,
doi: 10.1007/978-1-4020-6881-2_6, © Springer Science+Business Media B.V. 2008

or dose rate is expressed in watts per square metre (W/m^2), radiant exposure or dose in joules per square metre (J/m^2)

UV, especially UV-B, inhibits land plant growth but its effects are countered by DNA repair by an enzyme, photolyase, which craftily uses energy from visible light. The enzyme is found in all plants and animals except mammals, and there is some evidence that it was lost by placental animals, at any rate, 170 Myr ago.[1]

Plankton is also susceptible to UV damage, with serious implications for the marine food chain, but in some locations, such as the Sargasso Sea, it responds by generating dimethylsulphide (DMS), which promotes cloud formation and thus reduces the UV flux on the ocean surface. DMS concentration depends on the solar radiation dose received in the upper layers of the ocean irrespective of latitude, temperature or the concentration of chlorophyll-a, which is often used as a measure of biomass.[2] Radiant exposure or dose is expressed in joules per square metre (J/m^2), and irradiance or dose *rate* in watts per square metre (W/m^2).

The ozone and water vapour in the atmosphere, as we saw earlier, screen out UV radiation at wavelengths between 270 and 320 nm very effectively. UV-A increases its share from 76% to 96% by the time it arrives on the surface of the Earth at the expense of the rest of the UV wavelengths. UV-B (starting at 18%) is reduced to 4%. UV-C (starting at about 6%) is almost completely absorbed but it is at least 10 times more energetic than UV-A and UV-B and in parts of its range it may vary by a factor of two between solar minimum and solar maximum.[3] This makes UV-C relevant to long-distance airline passengers and to astronauts, just as it was on the early Earth when there was no ozone shield and as it has become in some parts of the world when the ozone shield has been damaged by our chemical profligacy. An interesting twist – one of many in the UV saga – is that UV-C lamps, which are advertised as germicidal, are extremely damaging to all biological tissue and the cornea in particular.[4]

The amount of sunlight reaching the individual thus depends on a host of obvious things: latitude, altitude, time of year, reflectance (albedo) of the ground, shading by topography and buildings, headgear and so on. The level of UV at noon may be ten times the level at 9 a.m. or 3 p.m. The main reason is the amount of scattering by the atmosphere, so that the Sun, which is white or pale yellow when overhead, often becomes a deeper yellow and then orange as the day progresses because its light has to traverse a greater length of atmosphere, and this scatters UV and blue light preferentially.[5] The day of the year is also important because the largest variation of UV irradiance (after clouds) is caused by the slant path of the radiation through the atmosphere.

Although UV-A and B can penetrate clouds, cloud cover may reduce net UV reaching the ground by over 50%. Thus August is much more dangerous than April in the northern hemisphere (the 2 months have approximately the same sun angle) because August usually has much less cloud cover than April, even though some cloud types increase the UV intensity on the ground by reflecting the sun's rays. Cold air can also be deceptive because temperature is not directly related to UV intensity, while reflective snow on the ground and high altitude raise UV levels significantly relative to low elevations, two items to be considered by skiers and climbers. The Global Solar UV index that is now provided daily by many authorities usually refers to conditions at solar noon, with 7–8 representing very high risk of sun damage and 10 min to sunburn for fair skins, but it cannot always take into account the effect of local elevation, which is commonly in the region of 10% per 1,000 m for UV-A and B.

Once again we are confronted by a paradox: ozone, our best shield against harmful radiation in the stratosphere, can inflame the lining of the lungs, aggravate emphysema, bronchitis and asthma, and reduce lung function. In the troposphere, or lower atmosphere, the atomic oxygen required to generate ozone is again formed by dissociation of oxygen atoms by sunlight but this time from nitrogen oxides and hydrocarbons rather than molecular oxygen. Raised levels of ozone can therefore result from industrial processes and agricultural practices that liberate methane such as cattle raising and rice growing in paddies. Ozone is also produced by some UV-C lamps.

The selective nature of the atmospheric UV filter is exploited by the Dobson spectrophotometer (hence Dobson units), introduced in the 1920s, which measures the amount of ozone in the atmosphere by comparing the amount of incoming UV light at 305 nm (which is absorbed by ozone) with that at 325 nm (which is not). Of course the wavelengths that are scattered by the atmosphere have already been distorted by the ozone in the stratosphere, but before the era of artificial satellites stratospheric ozone levels were difficult to evaluate. There was a suspicion that stratospheric ozone might be under threat from manmade pollutants, notably CFCs, but little importance was attached to the matter.

CFCs and bromofluorocarbons (or halons) were introduced in 1928 to replace toxic and corrosive compounds then used as refrigerants. Being chemically inert and non-flammable as well as non-toxic they soon found wide application as aerosol spray propellants and solvents for cleaning electronic printed circuits. One of the CFCs, CFC-11, was detected in the atmosphere at several locations over the Atlantic[6] but the concentration was 60 parts per trillion, 1/25,000 of the value for methane, and the CFCs seemed to pose no conceivable hazard. In the course of the 1970s, however, research showed that the very non-reactive properties of the CFCs would ensure their survival in the atmosphere for 50–200 years; and once in the stratosphere, under the action of UV radiation they would release their constituent chlorine, which would react with and destroy stratospheric ozone.

The resulting chlorine oxide soon breaks down, and a single chlorine atom can remain active long enough to destroy an estimated 100,000 ozone molecules. The process is helped by the development of polar stratospheric clouds in winter which provide particles on which stable chlorine compounds are converted into active chlorine. With the onset of spring, sunshine melts the particles to release the chlorine, which then drives the reactions. A similar but less emphatic process occurs over the Arctic, with bromine taking the place of chlorine.

Alarm at this threat led to a ban on CFCs in aerosol sprays in the USA, Canada and Scandinavia in 1976 but there was little sense of urgency as ozone depletion was estimated to total at most 3% a century. But in 1987 the Montreal Protocol, designed to protect the ozone layer by phasing out ozone-destroying gases, was signed by 43 countries. By 2007 the number of signatories had risen to 191. (The five exceptions included the Vatican.) The change of heart owed much to the discovery in 1985 by staff of the British Antarctic Survey that stratospheric ozone above the Halley Bay base had fallen to 30% of its usual value.[7] It is said that the ozone hole had been detected by high-altitude balloons and weather satellites in 1976 but rejected as implausible; at all events the results obtained at the research station were soon confirmed by data from the Total Ozone Mapping Spectrometer (TOMS) aboard the Nimbus-7 satellite.

Elsewhere on the globe the depletion amounts to about 5%. Between 1970 and 1980 UV-B levels reaching the Earth's surface in parts of Europe had increased by 10%.[8] Although by 1990 there was a noticeable slowing down in the rise of chlorine and bromine in the stratosphere, ozone mini-holes over Europe were still on the increase in 2006 and a European Commission report predicted that any recovery in the ozone layer could only become measurable around 2010 at the earliest. In June of the same year scientists from NASA and other agencies concluded that the ozone hole over the Antarctic (Fig. 6.1) would recover around 2068, nearly 20 years later than previously thought.

As with all such measures, the cost is more easily measured than the likely benefits. One estimate puts the cost of refitting and replacing industrial equipment that cannot accept CFC alternatives in the USA at $130 billion. If the CFC-ozone link proves false the outcome may be 'an unconscionable waste of resources, a loss of

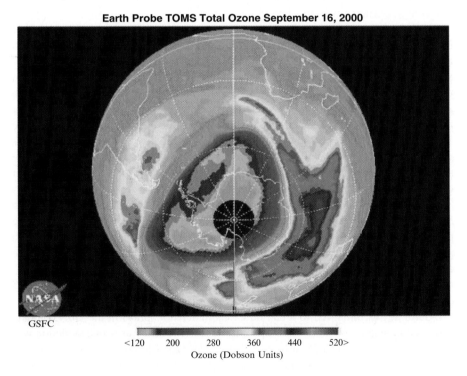

Fig. 6.1 The ozone hole over the Antarctic 15 September 2000 recorded by the Earth Probe Total Ozone Mapping Spectrometer TOMS (Courtesy of GSFC/NASA)

public trust, and a real setback for the environmental effort.' Besides natural variation, the sceptics point to ozone destruction by nucleation on particles derived from
volcanic eruptions, by water vapour or by air traffic, and they claim that there is no
evidence for an increase in UV radiation on the Earth's surface.[9]

Undoubtedly there are natural short-term fluctuations in ozone levels. At middle
latitudes the largest are driven by the Quasi Biennial Oscillation or QBO effect, where
winds in the Tropical stratosphere first blow eastward and then westward on an
approximately 2.3 year cycle[10] and cause ozone amounts at low and middle latitudes
to vary more markedly than during the solar cycle. The resulting changes in UV irradiance amount to ±15% at 300 nm and ±5% at 310 nm. A weaker ozone barrier
allows solar UV radiation to create more ozone in the atmosphere, which mitigates
the increase in UV, but this ozone has a relatively short life and thus does not make
up for ozone depletion in the stratosphere. The hydrochlorofluorocarbons (HCFCs)
by which CFCs have been replaced in some applications have the defect of being
powerful greenhouse gases. Their contribution to ozone destruction is estimated at
1/20 of that due to CFCs, and their production is not due to end until 2013.

Some cures are worse than the dangers they combat, said Seneca; yet inaction in
the face of persuasive evidence, especially inaction which sanctions suffering,
seems perverse.

SAD

Inadequate exposure to sunlight can lead to the mild but debilitating depression known
as seasonal affective disorder or SAD, an extreme form of the winter blues that is often
accompanied by low energy, weight gain, interrupted sleep and irritability.[11] The acronym is sometimes applied to social anxiety disorder. The two could well be linked.

Genetic factors may predispose some individuals to SAD either directly or
through variations in light tolerance.[12] A partial explanation for the affliction is that
the body's internal clock, which manages its circadian rhythms, is disrupted when
there is insufficient daytime light to switch off the production in the pineal gland of
melatonin, a hormone which induces sleep, and for the body to produce adequate
levels of the neurotransmitter serotonin, which promotes wakefulness.[13] However,
melatonin also influences the activity of the pituitary gland, which in turn controls
other glands responsible for the production of thyroid hormone, for example. And
serotonin has a strong effect on mood.

The switch is thought to operate by the passage of signals from the retina to the
suprachiasmatic nucleus, in an adjoining part of the brain, but there is some evidence that the bloodstream in other parts of the body which contain photoreceptors,
notably the back of the knee, which contains photoreceptors, can also mediate
between daylight and the biological clock. Whatever the route, any adaptive explanation faces the problem that the innate cycle dictated by the hypothalamus is
closer to 25 h than 24, whereas day length, as we saw, has if anything increased
(and very slowly) over time.

Since its first formal definition in 1948, SAD has been treated with bright actinic light. Indeed, there are reports to the effect that in vulnerable individuals it may be advantageous to have light-coloured eyes because they favour 'photic input', that is they are more responsive to phototherapy. Whatever the truth of this claim, phototherapy is also used to alleviate jet lag and other consequences of disrupted circadian rhythms such as fatigue during night shifts as the consequences can go beyond the inattention of businessmen: the accidents at the Three Mile Island and Chernobyl nuclear power plants and the Exxon Valdez oil spill have been blamed on shift-related fatigue.

Clinicians have concluded that the mechanism of light therapy is as elusive as treatment using antidepressant drugs.[14] That has not halted the growth of domestic phototherapy, and light boxes to be hung on walls or mounted on desks, like dawn simulators for the bedroom, are widely advertised. For some patients exposure to blue light (at 468 nm) has some beneficial effect[15]; others have benefited from cool-white or full spectrum fluorescent light. The patient sits in front of the box for a set period or has it targeted on his body from the side as he conducts his deskbound business. For greater mobility there is the sun-visor option, which is worn around the crown of the head.

With all these devices the UV component has to be strictly monitored. It has been suggested that UV-A might be permitted for SAD treatment because it raises levels of beta-endorphin and meta-enkephalin, opioids which influence feelings of well-being, but controlled studies show no significant increase in their level after exposure to UV-A,[16] and the risk of cancer cannot be justified.

Skin

No such controversy appears to surround UV-B phototherapy, sometimes limited to a narrow band within the UV-B range, in the treatment of severe psoriasis, acne and other skin diseases. It may also prove effective in controlling the symptoms of cutaneous lymphoma, as exposure to UV influences the metabolism and immune mechanisms of the skin. The most therapeutic wavelengths are 296–313 nm.

Conventional UV-B Broadband lamps cover this range and have been used successfully for many years. The risk of burning is minor, but photoaging of the skin can follow. Collagen is broken down faster than with chronological aging and the synthesis of new collagen is impaired. UV, whether in the clinic or in the Sun, particularly damages the loosely bonded long molecules of elastin, so that the skin no longer springs back. Damage accumulates, and freckles, age spots, wart-like actinic keratoses, fine wrinkles and a leathery texture may result.

UV-A treatment, if administered together with a psoralen drug such as methoxsalen, which makes the skin more sensitive to light, is known as PUVA. There are reports that some psoralens become active only when exposed to UV-A. PUVA treatment can be very effective against psoriasis although it brings the risk of burning and skin cancer. Psoralens in conjunction with UV were reportedly used in India in 1400 BC to treat vitiligo, in which the destruction of melanocytes, the cells that make pigment in the skin, leaves white patches on the body. The list of skin disorders which are improved by UV treatment includes some 30 items.[17]

Natural psoralens are found in a number of vegetables, including parsnips, celery, carrots and limes. Numerous forms of medication cause skin photosensitivity, including antimalarial, anti-inflammatory and anti-infective medication.[18] Extreme skin sensitivity to light, or photosensitivity, is an ailment in itself which affects patients with porphyria, for example, and commonly gives rise to a skin rash. The hereditary disease xeroderma pigmentosum is characterized by a sensitivity of the skin to sunlight which can lead to malignancies through complex toxic and allergic reactions (as in photodermatitis) as well as the failure of mechanisms of cellular repair discussed in the last section in this chapter.

Thousands of psoriasis sufferers visit the Dead Sea every year for climato-therapy. Partial or total improvement after four weeks of treatment is claimed for 80% of the patients and is attributed to the unique UV radiation environment.[19] The popularity of the Dead Sea among sunbathers of all sorts owes something to its position 420 m below sea level. Besides the longer optical path length traversed by the Sun's rays, additional scattering comes from the high atmospheric content in aerosols derived from the very salty Dead Sea; the dry climate appears to rule out any significant effect due to water vapour. More important, attenuation of UV-B is estimated to be almost 15% per 1,000 m whereas that for UV-A is just over 5% per 1,000 m. As noted later, UV-B appears the main culprit for sunburn, eye diseases and skin cancer, and the sunbathers appear to be justified in seeking the Dead Sea as a relatively safe location. Any benefits for the treatment of skin diseases would therefore seem to stem mainly from the reduced harm resulting from exposure to the Sun.

The message is in the dosage – and in the nature of the skin. As emphasised by the Cancer Research study that is cited at the head of this chapter, the essential precaution is to avoid burning. There is much evidence linking a small number of brief, high intensity exposures to UV and the eventual appearance of melanoma perhaps as much as 10–20 years later. Some Canadian studies claim that one or more blistering sunburns in childhood can double the risk of melanoma later in life; European studies reportedly state that three burning episodes in childhood increase by 60% the likelihood of developing melanomas in adulthood. Indeed much of the damage from photoaging is often achieved by the age of 20.

Penetration by UV into the skin depends on wavelength. UV-A can reach the subcutaneous tissue (although only 1% of the total radiation makes it there), whereas UV-C is confined to the outer skin or epidermis. The melanin particles in dark skin give optical protection to the cell nuclei. It stands to reason that some individuals are at greater risk of burning than others, but swarthiness can come with a genetic predisposition to skin cancer.

In the 18th and 19th centuries the moneyed classes coveted a pale skin, some would argue to distance themselves from outdoor labour, occasionally at the price of toxic lead-based cosmetics.[20] Whiteners contained substances such as zinc oxide, mercury, silver nitrate and acids as well as lead; some women even ate chalk or drank iodine to achieve whiteness.

Then, in the 1920s, a tan became desirable, the badge of increased leisure and foreign travel, or perhaps merely a shift in standards of beauty and of modesty combined with hearsay belief in the beneficial effects of sunshine. As we saw in Chapter 5, there is a good adaptive explanation for seasonal tanning at least among pale-skinned people living at high latitudes. But no such evolutionary forethought drives the acquisition of a permatan. The social historian Roy Porter[21] pointed the finger at Niels Finsen, whose researches 'inspired the unfortunate belief that suntans were healthy'.

Finsen was awarded the Nobel Prize in medicine or physiology in 1903 for his 'Finsen light therapy' for infectious diseases, especially lupus vulgaris, a form of tuberculosis of the skin. The Finsen Medical Light Institute in Copenhagen was founded in 1896; the first heliotherapy clinic opened in Switzerland in 1903. In the 1930s and 1940s the medical profession promoted sunbathing as beneficial for children and as a treatment for anemia, syphilis and kidney ailments.[22] Commercial tanning salons flourished.

The incidence rate of melanoma, which accounts for almost 75% of skin cancer deaths in the USA, had risen twentyfold between 1935 and 1996, and by 2004 was the fifth commonest cancer in men and the seventh in women in the USA. It is alleged that male teenagers are especially resistant to advice about sun exposure, and are accordingly punished in their 50s.

The effect of latitude on incidence is very pronounced: in New Zealand the incidence of melanoma is at least 37% higher in the north than in the south,[23] a latitude range of only 12°. In the USA, twice as many deaths due to melanomas are seen in the southern states of Texas and Florida as in the northern states of Wisconsin and Montana (a range of 20°). There are clearly many other factors involved besides latitude (Fig. 6.2), just as rednecks are not always sunburnt, but the Sun's elevation cannot be ignored.

The melanoma rate per 100,000 in Australia is three times higher than in the US.[24] It is intuitively to be expected that fair skinned Caucasians living in the Tropics should be at greatest risk; what is more surprising is that being born near the Equator brings a greater risk of developing skin cancer than emigrating there.[25] The same applies to Israeli-born Jews and those born elsewhere, perhaps because the effects of childhood exposure are indelible. Other risk factors include eye and hair colour, skin colour and ethnicity, hence the distinction often drawn between endogenous factors (genes) and exogenous factors of which UV radiation is one of several.

'Immediate tanning' is a darkening of the skin caused by oxidation of melanin under the action of UV-A. Delayed tanning, which occurs some 72 hours later, follows the production of new melanin in the skin by melanocytes in response to UV-B. UV-B also leads to thickening of the epidermis and thus provides some protection against UV damage. UV-A was formerly thought to be harmless but it is now known to promote suppression of the immune system as well as carcinogenesis[26] and cataracts. It penetrates deeply into the skin and into subcutaneous tissue, giving rise to long-term damage.

Sunburn can lead to skin cancer, of which the principal types are squamous cell carcinoma, basal cell carcinoma and melanoma (Fig. 6.3). The first two are usually

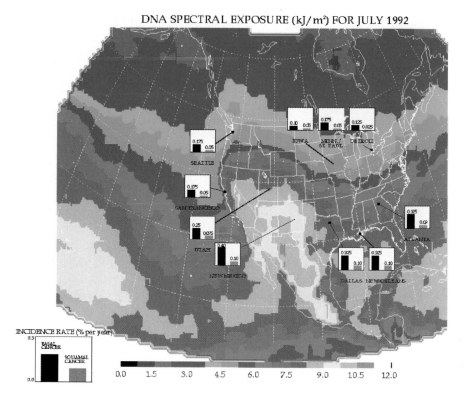

Fig. 6.2 DNA spectral exposure over North America for July 1992 determined from V7 Nimbus-7 data (Courtesy of NASA/GSFC). The scale at the bottom shows exposure in kilojoules per square metre, a measure of the total amount of UV radiation absorbed by the human skin during the day. The bar graphs show the incidence of basal (left) and squamal cancers (right). Note the tongue of high exposure over the western USA, mainly a product of elevation, and the incidence of basal cancer in New Mexico or Atlanta compared to Seattle or Iowa

grouped together as keratinocyte carcinoma (keratinocytes make up 90% of the cells in the epidermis). In 2004 there were over a million cases of squamous cell carcinoma in the US alone; the global figures show an annual rise of 3–10% in recent decades but its mortality rate is falling, in the US at any rate, thanks to earlier detection and treatment. The mortality rate of basal cell carcinoma, the commonest form of skin cancer, already very low, is also falling. Its incidence, however, has been rising by 20–80% according to location.[27]

Besides damaging skin, UV-B and UV-A break down water molecules to produce free radicals which are highly reactive and can injure tissue. Their effect can be countered by antioxidants, which include vitamins C and E and selenium. It has been estimated that every 1% decrease in the ozone layer should produce a 2% increase in UV-B irradiation which will lead to a 4% increase in basal carcinomas and a 6% increase in squamous-cell carcinomas.[28] Yet, as several observers have noted, the

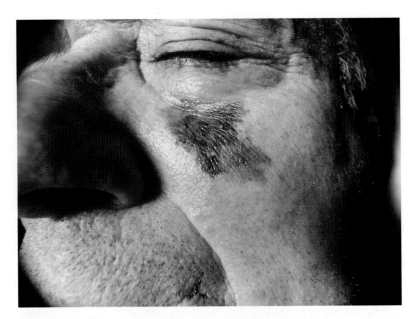

Fig. 6.3 Facial melanoma (Archive photograph from the Centro per lo Studio e la Prevenzione Oncologica of Florence, courtesy of Alessandra Chiarugi and Paolo Nardini)

current increase in skin cancer cannot be blamed on the ozone hole. The cause lies further back in time, and the most likely is the fashion for sunbathing and the greater opportunities to indulge in it, often at lower latitudes and higher altitudes, afforded by increased prosperity in the countries most affected. Those who dismiss concern over degradation of the ozone hole as premature would go further by arguing that only basal-cell and squamous-cell carcinomas are clearly linked to UV exposure and they are easily cured, whereas malignant melanomas are not linked to UV-B exposure but if anything to UV wavelengths above 320 nm which are not blocked by ozone.

The incidence of skin cancer is reduced by effective use of sunscreen during outdoors activity but this can be counterproductive if it is misleadingly reassuring. Sunscreen preparations are intended to reduce the amount of UV (and especially UV-B and UV-C, the main contributors to sun-induced erythema or inflammation) reaching the skin. They do so in two ways: mechanically and chemically. The first relies on very small particles of an inert substance, usually zinc oxide or titanium oxide, which scatter and reflect the UV energy. Chemical agents absorb UV and reradiate it at less dangerous wavelengths. Neither of them blocks the UV, whence the decision by the European Union in July 2007 to ban the terms *sunblock* and *100% sun protection* from sunscreen products. The only effective protection is provided by zinc oxide and clothing which is specifically designed to counter UV radiation. Rating of UV resistant clothing is in UPF numbers, which rate protection against both UV-A and UV-B; UPF 50 means that 1/50th of the UV radiation can reach the skin, and UPF 2 that a half will pass through.

Hitherto the emphasis has been on UV-B, and numerous sun screening agents are effective over the relevant wavelengths but some give rise to allergic reactions (such as contact dermatitis), stain clothing, are not water resistant, or are cosmetically defective. Although UV-A is now known to be a danger on some tanning beds it cannot be ignored in sunlight and there is increased emphasis on broad-spectrum preparations. The addition of avobenzone as well as titanium oxide and zinc oxide should provide adequate UV-A protection but how they are formulated and applied provide added complications. Some argue for oxybenzone and methoxycinnamate because they absorb UV-A. At all events sunscreen can be given a boost by taking vitamin E and selenium supplements and applying vitamins C and E to the skin.

Bone

UV-B is essential for the body's production of vitamin D3, which controls calcium absorption and helps to manage the movement of calcium between blood and bone. It plays an important role in regulating the growth of some 30 or more tissues, influences cell differentiation, maturation and death and switches genes on and off. It also acts as a hormone by altering growth signals to cells, inhibiting the growth of blood vessels – an important part of tumour formation – and modulating the immune system.[29]

Vitamin D is produced under the stimulus of UV-B from a substance (dehydrocholesterol) which is found in the skin and in the bloodstream and is processed into its active form in several different tissues of the body. Vitamin D deficiency can result in softening of the bones through defective mineralisation and sometimes termed rickets when manifested in children and osteomalacia for the adult version.

D. P. Hansemann (1858–1920) had drawn a connection between rickets and a lack of fresh air, sunlight and exercise, and later work showed that both sunshine and cod liver oil could cure it.[30] Adrian Palm in 1890 drew explicit attention to the value of sunlight in the treatment of rickets.[31] In 2007 *The Lancet* reported that rickets, 'once the scourge of Victorian Britain', was back. Some cases could be traced to macrobiotic diets which exclude such sources of vitamin D as dairy produce, meat and wheat but the commonest problem appears to be a failure to expose the skin to sunshine. UV-B is lacking in winter sunlight at high latitudes, and here the groups especially at risk include dark-skinned children and elderly people who do not venture outdoors. But the problem is not restricted to grey climates. It is common in India, Afghanistan and Australia, for example, where there is no shortage of sunlight.

One of the culprits is the veil or, worse, the burka, abetted by a predominantly indoor existence imposed on females and a diet which, through poverty or cultural bias, fails to make up the resulting vitamin D deficiency and may make things worse because a high intake of phytate derived from cereals inhibits calcification. The effect was noted as long ago as 1931 in Northern India. In Sydney a study covering the period 1993–2003 in three major hospitals[32] revealed a steady increase in rickets among children with a median age of 15 months of whom 22% already had

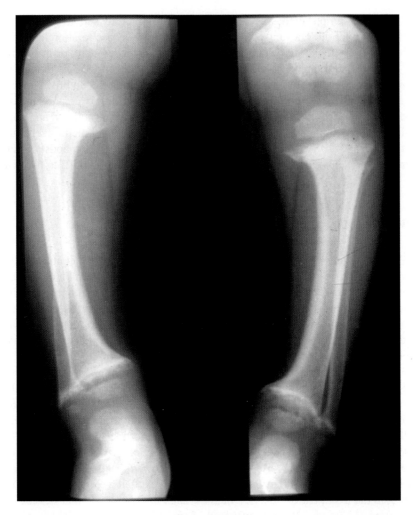

Fig. 6.4 X-radiograph of 3-year-old child of recent immigrants to the UK from India. Note bowed tibiae and widening of the low density zone below the zone of provisional calcification. The diagnosis is rickets due to vitamin D deficiency (From radiology data base at London South Bank University, courtesy of Ian Maddison)

bowed legs and 33% suffered from seizures due to low calcium levels in the blood. They were almost exclusively immigrant children or first generation breastfed children of immigrant parents from the Indian subcontinent, Africa and the Middle East (Fig. 6.4).

The vitamin is produced in sufficient quantity by the skin allegedly after as little as 15 min of daily exposure of the face and hands to sunlight, although the process appears to become less efficient with age. The optimum radiation wavelengths for the process are 295–300 nm, that is within the UV-B range. Window glass does not transmit wavelengths below 334 nm so that exposure to sunlight indoors is of no benefit for the production of D3 in the skin. Fluorescent tubes emit wavelengths shorter than 280 nm but according to some tests this is filtered out by the fixtures in which they are mounted. Websites are now available which explore the required exposure times for different locations and weather conditions.[33]

With increasing age the concentration of dehydrocholesterol in the skin decreases so that young adults can produce two to three times more vitamin D3 in their skin than elderly people. Melanin in dark skin absorbs some of the UV-B so that it requires longer exposure or higher intensity of solar radiation to produce the same level of vitamin D.

Some workers state that living as our ancestors did, in the open air near the Equator, we would synthesise 100 μg of the vitamin a day. The recommended daily allowance is 5 μg which rises to 15 if there is no exposure to sunlight. Vegans can obtain the vitamin from foods which are fortified by law in the UK, such as margarine and breakfast cereals, as well as vitamin supplements. Foods with naturally occurring vitamin D in the form of cholecalciferol are derived from animal products.

An additional benefit of modest exposure to UV-B is that the skin retains its ability to resist allergic reactions (or photodermatoses) provoked by exposure to the Sun. Most sunscreens let through sufficient UV-B for this and for adequate Vitamin-D synthesis in the skin but some dermatologists are so alarmed by the rapid increase in skin cancers that they are reluctant to recommend exposure to any UV radiation, and instead opt for supplemental vitamin D for instance from fish liver oil, which contains ergosterol (vitamin D2).

A countermovement has been launched by the respected medical journalist Oliver Gillie. His thesis is that the harmful effects of solar UV have been overstated, at any rate in northern industrial countries, and that vitamin D deficiency there increases the risk of contracting 16 several types of cancer, including those of the breast, bowel, ovary, and prostate, nervous system diseases including schizophrenia and multiple sclerosis, diabetes types 1 and 2, heart disease, high blood pressure, and schizophrenia as well as rickets and osteomalacia, and that it is a contributory cause of heart disease, raised blood pressure (hypertension), inflammatory bowel diseases, polycystic ovary disease, menstrual problems and infertility, infections and dental decay.[34]

In Gillie's view, a few minutes of daily exposure of the face and hands to sunlight is not enough. Only a minimum of three 20-min sunbathing sessions a week in bright midday sunshine will provide a white-skinned person with the requisite reserves for the autumn, winter and early spring when the sun is not strong enough to induce synthesis of vitamin D, whose levels in the body are halved roughly every 6 weeks. The need is all the greater in dark-skinned people, who require perhaps as much as twelve times as much exposure to the Sun to secure similar levels of vitamin D.[35]

Gillie acknowledges that excessive exposure to the sun may cause sunburn, skin aging, and skin cancer, although he thinks that perhaps fewer than half of the cases of melanoma skin cancer in these high latitudes may be attributed to sun exposure, and observes that, as vitamin D protects against melanoma, policies that discourage all exposure to the sun may be counterproductive. There could be no better illustration of the need to explain the complexities of the matter – the small print – to the potential beneficiaries or victims of sunshine.

In the USA the economic burdens of insufficient UV-B irradiation and vitamin D insufficiency from inadequate exposure to solar UV-B irradiance, diet, and supplements was estimated at $40–56 billion in 2004, whereas the economic burden for excess UV irradiance was estimated at $6–7 billion.[36] The authors of the study wisely add that further research is required to confirm these estimates. When considering the financial burden of a damaged ozone shield they will doubtless take into account the evidence that UV-B also damages plankton, and thus marine productivity as a whole. In one study carried out in 1992, Antarctic phytoplankton production was 6–12% lower beneath the ozone hole than outside it.[37]

Eyes

The tissues of the eye – the cornea, the aqueous humour in the anterior chamber, the lens, and the vitreous humour – transmit radiation to the retina, and damage may result from heating (photocoagulation) caused by visible and near IR radiation, which destroys a patch of rods and cones, as well as by the photochemical damage that results from exposure to intense light, for instance through unwise viewing of the Sun as a whole or during a partial or incomplete total eclipse. No pain is felt and the effect is not noticed until several hours later.

Not all wavelengths have free passage to the retina. The UV filter in the crystalline lens only protects the retina from UV-A (340–380 nm and 310–320 nm). The cornea absorbs all UV radiation with wavelengths <300 nm but high doses of UV can still cause the temporary clouding of the cornea called snowblindness. More acute effects include inflammation of the cornea and iris (photokeratitis) and an inflammation of the membrane that lines the inside of the eyelids (photoconjunctivitis). Long-term effects of UV exposure of the eye may include the development of pterygium, an opaque growth extending over the cornea which may cover the pupil, and squamous cell cancer of the conjunctiva. Pterygium is reportedly common in the southern and western USA as well as rural populations throughout the country; it affects 12% of Australian men over 60 and 33% of Chinese individuals over 50.[38] It may require surgery and can recur.

What is in effect a filter in the crystalline lens protects the retina from UV-A (340–380 nm and 310–320 nm). Transmission at 320 nm is present at birth but disappears with yellowing of the lens in the second decade.[39]

Cataract (Fig. 6.5) is perhaps the most distressing and publicized sunlight-mediated injury to the eye.[39] Yet the evidence for a direct link between UV and cataract is often fragile. For instance, a study in 14 of the states of the USA

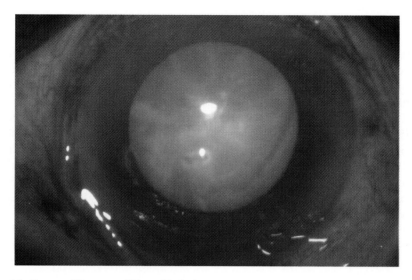

Fig. 6.5   Cataract (Courtesy of Pierre Hughet)

carried out in 1977 found that for patients aged over 65 the incidence of cataracts increased with hours of daylight; between 45–64 years the association was less clear and below 45 years it was slight. Among Australian aborigines it was correlated with annual UV-B level at their place of residence. But these and the many other published studies of UV and cataract found great difficulty in measuring exposure to UV because, as we saw, the UV dose is determined more from sunlight scattered by the ground and the horizon than from the sky. Moreover, besides age, other significant risk factors are diabetes, smoking and being female[40] and, a relative newcomer, corrective laser surgery.

Again, although some studies claim that a greater incidence of cataracts is found at high elevations, as in Tibet and Bolivia, and at lower latitudes near the equator, the contribution from environmental factors other than UV, let alone socioeconomic influences, is difficult to evaluate.

Yet the effort to refine the analyses is eminently worthwhile. Almost half the world's 36 million blind people are blind because of cataracts (and they would have included Galileo, who became blind at the age of 72 from a combination of cataracts and glaucoma) of which up to 20% may have resulted from exposure to UV radiation.[41] Increasing age is a key factor; in Australia, for example, the incidence of cataract doubles for every decade of age after 40. In India visually significant cataract occurs 14 years earlier than in a comparable group in the USA.

These problems will increase as the world population grows and ages. But, as the World Health Organisation[42] makes clear, age is shorthand for 'the complex interaction of exposure to many factors over time that contribute to the development of cataract'. And the two known effective ways to reduce the risk of cataract is to stop smoking and to reduce exposure of the eyes to UV-B radiation.

Protection against UV has to be complete for, as when observing eclipses, partial reduction in the light that reaches the eye may eliminate the squint reflex (squint in the sense of narrowing of the eyes rather than strabismus) and encourage the pupil to dilate, allowing more damaging exposure than might otherwise occur.

However, the major cause of reduced vision in Australia for people over the age of 55 is age-related macular degeneration. Laboratory experiments have shown that exposure to UV and intense violet/blue visible radiation is damaging to retinal tissue. It follows that effective protection requires sunglasses that block the appropriate UV wavelengths as well as the entire visual field especially when reflection from fresh snow (which reflects up to 85%) or water (up to 100%) is expected. General purpose sunglasses block 60–90% of visible light and UV-A and 95–99% of UV-B radiation. Special purpose sunglasses block up to 97% of visible light, up to 98.5%f UV-A rays, and at least 99% of UV-B rays. They are suitable for prolonged sun exposure but perhaps not ideal when driving.

Chapter 7
Space Weather

Cocooned in our geomagnetic bubble and bathed by the soothing solar wind we only dimly perceive the violence hourly unleashed by radiation, high energy particles and magnetic storms a mere 50 km above our heads and sometimes closer (Fig. 7.1).

This is the province of space weather, which in our usual self-centered way we define 'as the conditions on the Sun and in the solar wind, magnetosphere, ionosphere and thermosphere that can influence the performance and reliability of space-borne and ground-based technological systems and endanger human life or health'.[1] But, though unashamedly concerned with humanity's needs and mishaps, the science of space weather does not ignore grander issues; and, like a boat on a stormy sea, we provide some sort of scale to events which would otherwise prove unfathomable.

Our awareness of space weather began well before the space age. Perhaps the earliest hint of extraterrestrial disturbance came from the agitation of compass needles at times of exceptional auroras. By 1634 it was known that the Earth's magnetic declination – the difference between true and magnetic north – varied, and by the 1740s Anders Celsius in Uppsala had drawn a connection between magnetic and auroral activity. That it was a global effect was confirmed when compass needles flickered simultaneously in London and Uppsala.

These disturbances came to be known as magnetic storms because they resembled the effect of local lightning. In order to pursue the matter the explorer Alexander von Humboldt set up 'magnetic observatories' in France, Germany and Russia. In 1851 he publicized Schwabe's sunspot cycle in his *Kosmos*, a five-volume review of the state of science.[2] A year later Edward Sabine, Director of the British Colonial Observatories, reported that there was a correlation between the 1843–1848 sunspot sequence and the incidence of magnetic storms. The case was sufficiently persuasive for staff at the Greenwich Observatory in London to forecast in 1879 the next spell of magnetic storms at sunspot maximum.

For the next 50 years the role of the Sun as the fundamental cause of magnetic storms remained in dispute, and it took another half century for the association between large solar flares and severe storms to become clear.[3] This is understandable. Besides the need for adequate observational data, and therefore several decades of record, the physics had to progress beyond crude association, and much remained to be discovered about the various sources of geomagnetic storms on the Sun and

C. Vita-Finzi, *The Sun: A User's Manual*,
doi: 10.1007/978-1-4020-6881-2_7, © Springer Science+Business Media B.V. 2008

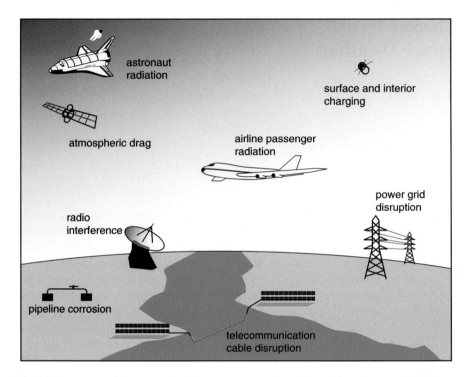

Fig. 7.1 Some potential targets of malignant space weather (Adapted from Lanzerotti 2001)

the complications introduced by variations in the Earth's field. The solar flare recorded in 1859 by Richard Carrington was followed 17 h later by unusually intense magnetic storms and auroras. No one appears to have read any significance in the delay except perhaps as evidence that the two events were unrelated. Although it was suggested as early as 1930 that magnetic storms represented plasma emitted by the Sun, the physical fact of mass ejections remained difficult to grasp without adequate imagery and the continuous, unobstructed observation provided by artificial satellites.

Even then the evidence remains largely circumstantial, and the Sun must be considered innocent of causing the turmoil until demonstrably guilty: other unusual suspects include stars in their death throes, within our galaxy or in a distant one, capable of accelerating particles to the status of GCRs.

As it happens, there long prevailed a distinct reluctance to indict the Sun even when, as the Director of Kew Observatory delicately put it, in 1859 'our luminary [was]… taken in the act' of producing a magnetic response.[4] Its earliest manifestations went unrecognised partly because interplanetary space was considered a vacuum incapable of transmitting energy. Now there are flotillas of spacecraft and countless bugging devices hoping to document, and thus perhaps predict, mishaps and disasters; and space is seen to be awash with plasma.

Communications

The heliograph, a mirror used to transmit messages by reflecting flashes of sunlight, doubtless served Antiquity well and is still of service in emergencies today. In 1880 Alexander Graham Bell, inventor of the telephone, created the photophone, which transmitted the vibrations of a mirror into which the sender spoke to crystalline selenium cells at the focal point of a parabolic mirror serving as receiver using a beam of sunlight as carrier.[5] Its maximum range was just over 200 m, and transmission was disrupted by clouds and other nuisances, but it ranks as a remarkable precursor of optical fibre transmission.

Otherwise the Sun has contributed little to the technology of communications and if anything, has tended to provide an array of obstacles. Magnetic storms, in particular, were an inconvenience and doubtless an occasional hazard to compass-using navigators, and their negative impact on human activity grew dramatically with the development of the electric telegraph.

Surprisingly, no one seems to have a bad word for the aurora, that shimmering curtain of light hauntingly flitting across the polar sky. As we saw in Chapter 3, both the northern and southern varieties – borealis and australis – signal the collision of the solar wind with nitrogen and oxygen in the upper atmosphere. Superstitious or psychological side effects apart, it seems a harmless enough spectacle. But during spectacular auroras in 1872, 1882 and 1894 telegraphs were disabled in many countries.[6]

> The telluric currents attained an extraordinary development during the aurora of February 4, 1872, ...one of the most extensive known; it was seen in the whole of the West of Asia, in the North of Africa, throughout Europe, and on the Atlantic as far as Florida and Greenland; at the same time an aurora was observed in part of the southern hemisphere. The disturbances in telegraphic communications were no less extensive.[7]

By 1940 telephone networks were also being perturbed during auroral displays.[8] The wireless telegraph, first demonstrated in 1901, was not immune from solar interference as it depended in one way or another on the ionosphere, the name given to plasma in the thermosphere produced mainly by UV and X-ray radiation from the Sun to different extents according to the season and to the time of day and night.

In the early decades of wireless radio, long-wave transmission was the rule, and its signals could travel long distances thanks to the waveguide effect provided by the lowest part of the ionosphere – the D layer – and the Earth's surface. When cheaper shortwave systems began to dominate radio communication their signals followed the Earth's curvature by refraction at the base of the ionosphere.

Guglielmo Marconi did not fail to notice that signal fading practically always coincided with large sunspots, intense auroras and magnetic storms,[9] but the long-wave techniques on which he initially depended were relatively insensitive to such interference. Short-wave frequencies fare more or less badly according to position in the solar cycle because the efficacy of refraction depends on the degree of ionization produced by solar UV radiation. On 25 March 1940, all short-wave contact between Europe and the USA was interrupted during a solar storm; on 2 February 1946 widespread radio interference coincided with naked-eye observation of a large sunspot group.[10] When the 1944 code-breaking computer Colossus was rebuilt at Bletchley Park, in England, the first test message sent from Germany in November 2007 was made unreadable by solar interference. As the engineers in charge stated: 'this is the worst point in the 10-years sun cycle' to make the attempt. 'It happened often in the war....'[11]

Artificial satellites have eliminated the uncertainty created by variable ionization by providing a dependable refracting surface. But they are subject to solar interference as regards their elevation, their electronics and their radio links: true creatures of the space age. Indeed there are settings where their dominance has already passed. In 1988 satellites carried most conversations and other data across the oceans, as they were deemed more cost-effective and flexible than cable systems on land and under water for distances of over about 700 km[12]; by the year 2000 over 80% of the traffic had been lost to submarine cables enjoying the wide bandwidths, rapid response time and greater security made possible by fibre optics.[13]

The fibres consist of very pure glass. Single-mode fibres have a diameter of about 8 μm and transmit IR laser light (1,300–1,550 nm); multi-mode fibres have a diameter of about 62.5 μm and transmit IR light (850–1,300 nm) from light-emitting diodes (LEDs). As well as being cheap, light and thin, optical fibres can handle a large volume of traffic with little interference. By 1992 the version known as 2-9, which linked North America with Europe, could carry 80,000 simultaneous conversations over a distance of more than 9,000 km, twice the capacity of TAT-8 installed 4 years earlier. TAT-14, introduced in 2001, has a capacity over 2,000 times greater.

Yet the long arm of solar interference reaches even the seafloor. There had been disruption of communication on submarine cables as early as 1872 from Lisbon to Gibraltar, along the Mediterranean, from Suez via Aden to Bombay, and across the Atlantic. In March 1940 the transatlantic cable between Scotland and Newfoundland experienced surges of up to 2,600 V. In February 1958 a large potential difference disrupted TAT-1 between Newfoundland and Scotland.[14] The effect arises when electrical currents are induced in the Earth by time variations in the geomagnetic field caused by sudden changes in the solar wind.

The problem has gained in significance from the transfer of most of the transoceanic traffic in messages and data from satellite to fibre. Despite their reliance on light the optical cables still embody a conductor for the repeaters that are required every 100 km or so to offset losses in transmission. Technical improvements will reduce but are unlikely to eliminate completely this avenue for induced voltage.

Satellites

The growth in submarine traffic did not result in empty skies. In 2007 there were an estimated 850 civil and military satellites in Earth orbit of which some 560 (66%) were devoted to communications.[15]

In March 1989, when the northern aurora extended as far south as the Mediterranean, the orbits of numerous satellites were disturbed. Satellite failures were reported in 1991, 1994 and 1997. In 1998 a solar storm disrupted a Galaxy IV satellite which supported automatic cash machines and airline tracking systems. On 28 October 2003 another storm damaged one of the instruments aboard NASA's 2001 Mars Odyssey orbiter as well as two Japanese satellites. The setbacks are not confined to electronics. Satellites are sensitive to atmospheric drag. Heating of the upper atmosphere is of course greatest at solar maximum and can raise the temperature of the upper thermosphere to 2,500°C or more. As the atmosphere as a whole expands, lower and cooler portions rise into the path of the spacecraft and the local air density increases as much as fiftyfold, whereupon drag on the satellite is enhanced and its orbit decays more rapidly than expected.

Related effects that are especially critical for low Earth orbit satellites – those that operate 200–2,000 km above sea level – are even more spectacular. The concentration of atomic oxygen (that is to say oxygen not incorporated in the O_2 molecule) at elevations of 500–800 km can vary a thousandfold over the solar cycle; at high concentrations the oxygen may react with satellite surfaces and instruments and cause physical damage.[16]

The best known example of orbital degradation is Skylab I, the USA's first orbiting space station (see Chapter 1), which carried, among other instruments, the Apollo Telescope Mount to be used for spectrographic analyses of the Sun without interference from Earth's atmosphere. The first of three crews to man the craft successfully repaired its sunshade and a solar array, both broken during launch, and several solar experiments were conducted during its useful year of life during which coronal holes were discovered. Mainly in response to increased solar activity – solar Cycle 21 was to peak in December 1979 – Skylab I fell to Earth and broke up after 6 years in operation on 11 July 1979. The Solar Maximum Mission satellite began its life as a low Earth orbit satellite. It was repaired by the crew of the Space Shuttle in 1984 but an attempt to boost its orbit failed and it reentered the atmosphere in 1989 disappointingly just as solar activity was picking up.[17]

Some satellites, such as the Tropical Rainfall Measuring Mission, have onboard jets to compensate for orbital decay. The Hubble Space Telescope does not, and it has to be periodically manhandled into a better orbit by astronauts ferried up by the Shuttle. Orbital 're-boosting' during benign periods of solar activity helps to prolong the desirable HST orbit by approximately 1–2 years, whereas re-boost during the solar cycle's climb to its peak activity can be beneficial for as much as an entire decade. If successful, the last servicing mission in September 2008 should extend the life of the HST to somewhere between 2022 and 2028. The hope is that after the Shuttles are grounded in 2010 some other mode of access will become possible.

The only beneficial effect of atmospheric drag, in the eyes of NASA, is removing some of the junk from space. According to the Goddard Space Flight Center there were 6,133 bits of unwanted debris in Earth orbit in April 2000. The gain in tidyness is partly offset by the difficulty of tracking these objects as they reenter the atmosphere.

Besides drag there is the problem of charging. During magnetic storms, satellite electronics can be damaged through the build up and subsequent discharge of static electricity. Electrical discharges may eventually arc across spacecraft components, harming and possibly disabling them. Bulk charging occurs when energetic particles, primarily electrons, penetrate the outer covering of a satellite and deposit their charge in its internal parts. If sufficient charge accumulates in any one device, it may discharge to other components. Radiation can also jeopardise an entire mission by directly damaging the solar cells that power the control and data-gathering systems.

Satellites are perhaps most familiar in everyday life through their role in GPS, the US Defense Agency's Global Positioning System. (Competing systems are planned by Europe, China, Russia and India.). Initially the quality of its results for civilian users was purposely degraded but in May 2000 this 'selective availability' was abandoned by the US Government. Some reports say this was because a Korean airliner which was shot down over Soviet territory in 1983 had wandered off course; others claim it was it was because scientists had found a way of getting round the restrictions.[18] The GPS system employs at least 24 satellites at an elevation of 20,200 km and transmitting microwave signals which make it possible to compute horizontal position with an accuracy of a few metres, even with handheld devices, when at least four satellites are in optimal positions aloft. With prolonged observation using two instruments of which one is at a known location ('differential GPS'), accuracies of a few millimetres are attainable.

The relativity effect could result (as noted in Chapter 2) in errors of up to 10 km a day. General relativity predicts that clocks will run faster in a weaker gravitational field than those in a stronger field; special relativity that moving clocks will run more slowly than stationary ones. The former effect means that clocks in a GPS satellite moving at 3.9 km/s will run 6,200 ns per day slow; the latter, that they will run 45,900 ns/day fast.[19] The net effect can be corrected in advance by adjusting the rate at which the clocks run. The requisite difference is 10.2299999995453 MHz compared with 10.23 MHz.

The main sources of error, besides those arising from malfunction of atomic clocks and computing devices, reflection of signals from obstacles ('multipath'), and human error, are those due to disturbance of the satellite orbits and changes in the speed and direction of the radio signals. Orbits are affected by gravitational forces, including those due to the Sun and Moon, by the solar wind, by the Earth's albedo and by differential radiation from the satellite itself. The signals may undergo refraction in the ionosphere, which is of course affected by the solar wind and which has a different effect on high and low frequency signals, and also refraction by water vapour in the troposphere.

If the changes are smooth they can generally be countered during or after the time of observation. Solar storms present serious problems because they are difficult to model, witness the difficulties encountered during an attempt to use GPS for measuring vertical ground motions in Iceland. A network focusing on eastern Iceland had been set up in 1987 and resurveyed in 1990, 1992 and 1995. Only the 1987–1992 epoch yielded a statistically significant vertical deformation field as the

1990 data were collected at the peak of a sunspot cycle and consequently degraded by ionospheric noise.[20]

Charged particles from solar flares also produce intense bursts of radio noise, which peak in the 1.2 and 1.6 GHz bands used by GPS. Normally, radio noise in these bands is very low, so that receivers can easily pick up weak signals from orbiting satellites. It is claimed that the problem was not spotted until recently because the use of GPS systems blossomed during a period of relatively low solar activity[21] and thus when there were few solar flares.

This accords with the events of 5 and 6 December 2006, when forecasters from the National Oceanic and Atmospheric Administration (NOAA) observed two powerful solar flares which originated from a large sunspot cluster. The flares gave rise to a solar burst which, though it occurred during solar minimum, produced ten times more radio noise than the previous record and, at its peak, 20,000 times more radio emission than the rest of the Sun: enough to swamp GPS receivers over the entire sunlit side of the Earth. But the flares in October and November 2003 (the 'Hallowe'en flares'), also near solar minimum, were even more spectacular. It will be recalled that the X-class, the highest category, ranges from 1 to 9; the flare on 28 October measured about X28. Unusually, but not surprisingly, it was accompanied by a perceptible increase in total solar irradiance.[22]

Power grids and pipelines

During solar storms, the solar wind subjects the Earth's magnetic field to severe fluctuations, and potential differences in voltage are induced between grounding points in electric grids much as at the extremities of transatlantic cables. The resulting currents may cause saturation of transformers and shorten their operational life, and disrupt the operation of relays and circuit breakers leading to systems shutdown.[23] The risks are enhanced where, as in northern North America, much of the electrical grid is near the North magnetic pole and the 'auroral electrojet current', an oval zone around the North pole in which flow currents attaining several million amperes. The predominance of igneous rock, which commonly has high electrical resistance, encourages any geomagnetically induced currents to flow in the transmission lines rather than through the ground.[24]

A solar storm thus disrupted electricity supplies in parts of the USA and Canada in 1940. In 1972 a 230,000 V transformer at the British Columbia Hydro and Power Authority exploded. In 1958 there was a power failure in the Toronto area during which the only light was provided by the aurora.[25] In March 1989 a solar storm led to a blackout in Quebec which lasted 9 h and affected more than 6 million customers.

These events cannot match in scale a blackout in 1965, which affected 25 million people for up to 12 h over Ontario and the northeastern USA, the 1999 blackouts that left 90 million people without power for several hours in Brazil and 8.5 million consumers in Taiwan, or the 2003 blackouts that again affected about 50 million people in Ontario and parts of northeastern and midwestern USA. All, so far as we know, were due to hurricanes, landslides, human error and other terrestrial agencies. Al Qaeda claimed responsibility for the North American 2003 event but a report released in 2004 blamed the failure on contact between overloaded high-voltage power lines and trees which had been inadequately trimmed,[26] and the consensus now favours a simple case of overload. But all the above demonstrate how failure can cascade through extensive networks. The 800,000 km of bulk transmission lines in the USA and 12,000 major substations offer numerous entry points for geomagnetically induced currents, and power grids operate with little spare capacity. The giant power grids being proposed for the Mediterranean and North Africa and discussed in Chapter 8 offer even more scope for induced mayhem.

The impact of storms on power systems can be mitigated by reducing current flow with series capacitors in transmission lines or in the earth wires of transformers. Several such have been put in place in Quebec since 1989 but they are expensive and their installation is not straightforward. In the short term, effective forecasting might allow timely reduction in loading of the system to introduce more leeway and in other ways ensure that operators have emergency procedures in place.[27] But restoring power after an extensive blackout is perhaps more problematic especially as demand can then be six times the normal level. Equipment is damaged, delaying restoration further. Large transformers can cost over $10 million each and may take over a year to replace. The cost of a major blackout in France lasting 4 h was put at a billion dollars; a major blackout in the northeastern USA would now result in losses measured in several billions of dollars; extrapolation to the flare of 1859 suggests that the related blackout could affect half the population of the USA, with losses estimated in trillions of dollars.

The issue is not simply one of industrial and domestic power, and the delay its restoration entails, but also of the many services whose disruption can have serious consequences, such as water supply, sewage treatment and disposal, refrigeration, public transport, airline management, and banking.[28] Induced currents are hosted by long, metallic, grounded water, oil and gas pipelines, especially those that extend into auroral regions and their periphery.[29] A power gas pipeline explosion in June 1989 caused by leakage from a corroded section engulfed two trains on the Trans-Siberian railroad, with hundreds of deaths. By 1977 the Alaska pipeline already measured 1,300 km from 62° to 69° N within the most active part of the electrojet current mentioned above; modelling of the induced fields during several solar cycles showed that half the time the induced currents would be less than 1 A but during high solar activity surges could exceed 500 A.[30]

The currents enhance corrosion by a process analogous to the deterioration in the zinc casing of a dry cell. They flow from anodic areas on the pipeline to cathodic areas through the soil, which acts as an electrolyte, and corrosion occurs at the anodic area. Damage can therefore be reduced by coating the pipe with electrically insulating material, by maintaining the pipe at a negative electrical potential relative to the ground to inhibit oxidation of the pipeline, or both, simple but expensive expedients.

A more complex version of the cathodic technique follows the discovery in 1824 by Humphry Davy, inventor of the miner's safety lamp, that the corrosion of ships' copper hulls was reduced if lumps of iron were attached to the ship below the waterline. The essence of the technique is to make the surface of the pipe serve as cathode of an electrochemical (galvanic) circuit and, as in Davy's prototype, to introduce a 'sacrificial' metal. It is widely used for well casings as well as pipelines of all kinds in a wide range of environments.

The erratic nature of the field fluctuations during a magnetic storm could undermine any such electrical protective measures by creating surges that exceed the limits of cathodic protection. Moreover, according to some authorities, modern pipe coatings make things worse by creating greater pipe-to-soil potentials than in the past so that any defects in the coating can increase the risk of corrosion.[31] The most effective (and expensive) solution remains to insulate the pipe more thoroughly from the ground whether below the surface or above it.

The human target

The supposed dangers to which humans are exposed in Space was vividly conveyed by the cover story of the tabloid *France Dimanche* of 22 July 1969, which revealed **The terrible admission made by the Apollo astronauts to their wives,**[32] evidently not talking about their powers of concentration or their digestive systems.

Outside the bulk of the Earth's atmosphere the risks rise dramatically. Some sources firmly state that the most deleterious factor of space is solar UV radiation[33] but it is relatively easy to ensure as much protection from UV as is provided by the atmosphere. Not so when it comes to radiation from X-rays and gamma rays.

The rad is a unit equivalent to the delivery of 1 W of energy into 100 kg of human tissue. Thousand rad will bring death with a few days; 0–50 has no obvious short-term effects. As there are many sources of radiation, we also use the rem to denote dosage. Thus 1 rad represents 1 rem of X-rays or gamma rays but 20 rem of alpha particles. In the USA the annual background exposure including that derived from food averages about 210 mrem. Cosmic rays add another 100, giving an annual average of 300 mrem. The US Occupational Safety and Health Association stipulates that astronauts may not exceed 50 rem above their normal background a year or 25 rem in any 30-day period or 100–600 rem for their career lifetime dosage.[34]

Astronaut Shannon Lucid reported that, on the Russian Mir space station, the typical radiation dosage was the equivalent of about eight daily chest X-rays.[35] During a storm at the end of 1990, Mir cosmonauts received a full year's dosage in a few days.

The calculation is not a straightforward one. In September 2005 an active solar region spouted solar flares but radiation levels in the International Space Station dropped, with GCR levels 30% lower than usual. This was because the CMEs and the magnetic fields they conveyed swept aside charged cosmic ray particles. The protons accelerated by CMEs can also produce radiation sickness but few solar protons can penetrate the walls of NASA spaceships. Thus a craftily scheduled space mission can use solar maximum to reduce the GCR flux at dangerous levels by 30% or more.

As regards the Apollo missions, when such scheduling was perhaps not feasible, NASA gave this reassuring answer to an Internet enquiry which asked, among other things, *Why were the astronauts, who traveled to the Moon, not killed by exposure to radiation?*

Astronauts on Apollo missions were exposed to radiation from the Van Allen belts (trapped particles) and the galactic cosmic ray (GCR) background. There were no solar particle events of any significance during these missions; hence, there were no dose accumulations from them. Radiation exposures for the transits through the trapped belts were kept very low by transiting quickly through them. Since the Apollo mission lengths were only a week or two, there were also no significant accumulations of radiation dose from the GCR background. Thus doses were well below thresholds for any measurable radiation effects, including lethality.

According to some reports, NASA instructed its astronauts to rotate the space capsule in order to place the fuel tank as shield between them and the Sun if a flare was expected.

The Apollo astronauts absorbed perhaps the level experienced by Shannon Lucid on the International Space Station, 400 km above the Earth and shielded in part by the Earth.[36] But it was a close run thing. A National Research Council report on space radiation hazards found that on 4 August 1972, 3 months after the return to Earth of the crew of Apollo 16, the largest Solar Energetic Particles event of Solar Cycle 22 began to launch high energy protons. They arrived at Earth 40 min after a large solar flare was sighted. Estimates of the dosage they would have inflicted on the astronauts in their spacesuits and perhaps even in their cabin suggest it was sufficient to produce severe skin damage.[37]

The evidence from ice cores indicates that there have been some much larger events during the last few centuries, implying potentially lethal dosages about four times as great as those in 1972. A recent catalogue[38] of polar and Greenland records indicates 154 large solar particle events in the period 1561–1992; the largest spike corresponds to the 1859 flare. The available data suggest that, during a Carrington-type event, astronauts behind only a few centimetres of aluminium – the shielding found in the average spacecraft – would suffer a dose that could cause acute radiation sickness and possibly even death. Parts of the International Space Station, however, have enough shielding even for this worst-case event, and if the Sun's activity is closely monitored the crew should receive several hours' warning to shelter from a large flare.

But there are implications for the design of long-duration interplanetary manned missions. Conventional aluminium might not be the best choice of building material for a spacecraft for Mars when, in any case, the greatest threat is likely to come from GCRs, which have been greatly accelerated by supernovae to energies perhaps 10,000 times greater than typical particles accelerated by solar flares, as they can readily damage cells and DNA.[39]

Even airlines take discreet note of the problem. United Airlines, and doubtless others, reroute polar flights to avoid significant solar events in the knowledge that inflight radiation doubles every 2,000 m of altitude. At 10,000 m the normal dosage rate might be 0.6 mrem/h. According to the Health Physics Society the 40–60 mrem that a severe solar flare may give is about what is received during a chest X-ray and far less than a CT scan.

Other commentators are less sanguine. For example, a study conducted for the World Health Organization concluded that airline crew members receive radiation doses greater than that of a typical nuclear energy worker, and recommended the wearing of dosimeters.[40] European Union legislation introduced in 2000 is more permissive and requires the member States to 'take into account' assessed exposure to ionizing radiation when organizing work schedules, to keep the workers informed of the risks they face and, by applying Article 10 to female aircrew, to treat the unborn child as a member of the public. That might not reassure passengers but being cheerfully patronised does not either.

Future developments will cut both ways: faster travel will reduce transit time but higher altitudes will increase route doses, and however rapid the journey there remains the risk of encountering a shortlived major event such as a solar particle event created by a solar flare.

Forecasting

Everything that has gone before, and not just electricity supplies, would benefit from advance warning: the science, the engineering, and the travel if only to the extent that one could quantify one's anxiety. Quite apart from storm-related hazards, predictions of solar and geomagnetic activity influence the efficient operation of orbiting satellites, geophysical exploration, and high-frequency radio communications and radars.

The most basic mode of forecasting depends on the 11-year sunspot cycle. Enough data have been gathered for effective analysis of the timing and probable magnitude of the next peak. The first broadcasts of geomagnetic data were made from Paris in 1928. Since then there has developed a global network of solar observatories which pool data gathered from the ground and from space. The international sunspot number is produced by the Solar Influences Data Analysis Center (SIDC), part of the World Data Center for the Sunspot Index at the Royal Observatory of Belgium. National bodies, such as the National Physical Laboratory of India, also provide sunspot predictions 6 months in advance for home use mainly for planning High Frequency communications and for managing satellites in low Earth orbit in the context of changing atmospheric drag. Data published on the Internet also include solar flux values corrected to what they would be at 1 AU so that different locations at different times can be sensibly compared.

The sunspot cycle provides a broad framework for crude estimates of phenomena related to levels of solar activity (Fig. 7.2). CMEs, as noted earlier, occur 15–20 times more frequently at solar maximum than at solar minimum; flares, on the other hand, can be at their most devastating near solar minimum; radio emission at 10.7 cm (2.8 GHz), which has been measured daily since 1947 and is a good guide to

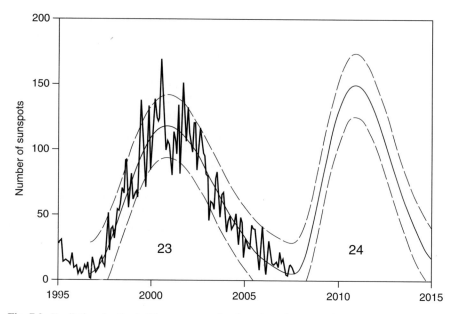

Fig. 7.2 Prediction for Cycle 24 sunspot number based on data up to 31 March 2006 (Courtesy NASA, MSFC and David Hathaway). The intensity of the new cycle is in dispute, with estimates for the peak, expected in 2011--2012, ranging between a moderately strong 140±20 sunspots and a moderately weak 90±10 sunspots

the Sun's UV output – significant in stratospheric heating and thus is assessing satellite drag – closely follows the sunspot curve and can be estimated from it.

The basis for prediction is thus still largely empirical. One of the methods used by NOAA combines regression and curve fitting, which work well as solar activity approaches a peak, with information from precursors such as measurements of the Earth's magnetic field, which is more informative near solar minimum.[41] Radiocarbon evidence has been used to reconstruct sunspot numbers back to 8505 BC; the pattern that emerged was the basis for prediction of solar activity[42] from the present day as far ahead as AD 2045. The results indicated that solar activity will on average be weaker during the first decades of the 21st century than it was during the last decades of the 20th. Though very tentative this kind of analysis is clearly important when the solar factor in climate change comes to be assessed.

A 12-member panel at the US Space Environment Center in Boulder, Colorado, was evenly split over whether the next peak would measure 90 or 140 sunspots, although there was agreement that it would come in October 2011 or August 2012. The difference over magnitude is slight as an average maximum ranges between 70 and 155 sunspots. Timing, however, is crucial, as precautionary measures are generally most effective if both the date and the likely duration of the problem period can be specified. For example, the estimated value of satellites susceptible to damage is over $200 billion, and floating oil

rigs depend on GPS to remain within a few centimetres of their preferred location if drilling gear is not to be damaged.[43] As increased drag in the upper
atmosphere affects space debris as well as satellites it needs to be monitored to
ensure the safety of space flight. The great increase in the range and number
of devices sensitive to space climate means that problems that once arose only
during solar superstorms have been rendered commonplace.

This is clearly the case with regard to the impact of space weather on commercial
airlines, as it goes beyond the issue of passenger exposure to radiation to include
avionics, communications and GPS navigation devices.[44] Increasingly sensitive electronic devices may be damaged by cosmic rays, solar particles and secondary particles
produced in the atmosphere, resulting in corrupted memories, erroneous commands
and even hardware failure. The problems will inevitably increase in importance as the
number of long-distance flights at high elevations and dangerous latitudes increases.

A looser but still empirical approach governs attempts to forecast the impact on
the Earth of events observed on the Sun. The interval between a flare and the arrival
of potentially dangerous solar energetic particles (SEPs) may be a few hours, and
between the launch of a CME and the arrival of the shock wave it creates may amount
to a few days, sound reasons for perpetuating the kind of continuous monitoring of
the Sun being performed by SOHO. In fact, the intensity and rate of an initial electron
surge observed by SOHO has made it possible to predict the intensity of the more
dangerous, heavier ions with a lead time of up to 74 min. Tested retrospectively for
2003 the method correctly predicted the four major ion storms of that year. Granted
that it raised three false alarms the method clearly has great potential for alerting lunar
astronauts (and satellite operators) that a dangerous storm is on its way.[45]

The Australian Space Weather Agency issues online information on solar conditions – solar wind speed, X-ray flux, X-ray flares – and from Culgoora observatory
real-time spectrographs and H-alpha and visible light images of the Sun.[46] In 2000
NOAA introduced a set of Space Weather Scales designed to 'communicate to the
general public the current and future space weather conditions and their possible
effects on people and systems.' The scales describe the environmental disturbances to
be expected from three event types: geomagnetic storms, solar radiation storms, and
radio blackouts. They are all rather narrowly defined, doubtless to avoid academic
waffle. Geomagnetic storms range from G5 Extreme through Severe, strong and
Moderate to Minor. Solar radiation storms and radio blackouts likewise range from
extreme to minor. The solar wind data (velocity and proton density) are updated every
10 min. They are derived from real-time information transmitted to Earth from the
ACE spacecraft which, being located at the L1 libration point between the Earth and
the Sun, can give about 1 h advance warning of impending geomagnetic activity.

Solar weather forecasts can be expected to be as routinely available as atmospheric weather forecasts. But they will require a large number of expensive, strate-
gically placed satellites working in unison with ground-based observatories which
are manned by experienced observers who can make sense of what are still often
ambiguous signals. It is some consolation that some of the events we currently find
hardest predict, notably solar flares, appear to create serious damage in radio communications more readily than in human tissue.

Chapter 8
Solar Energy

Enthusiasts for solar power need to be reminded that, through its role in photosynthesis and in powering the atmosphere, the Sun is already our primary source of
renewable energy. Or to put it another way solar photons convert naturally into
chemical fuel and heat.[1] Photosynthesis yields biomass which can serve as fuel and
food and it creates ATP, the main source of energy for living things and thus their
muscles. In the process it liberates the oxygen required by most kinds of combustion Solar heat sets up the differences in temperature and pressure that propel wind
and waves; and by providing rainfall and melting snow it generates the mechanical
energy that drives water mills and their hydroelectric successors.

The only significant non-solar sources of energy are the tides (the Sun, though
undoubtedly much larger than the Moon, is so distant that its pull is only half as
strong as the Moon's), geothermal heat – heat from the Earth's interior – and
nuclear power. An estimate by the International Energy Agency put their respective
contributions to global energy consumption in 2006 at a pathetic 0.0004% for tidal
power, 0.06% for wind and a thousand times more but still only 0.4% for geothermal. Nuclear accounted for 6.5%.

Yet as a *direct* source of energy the Sun contributed only 0.04%.The figure probably underestimates current use, especially as many consumers of solar energy are
found in isolated locations, such as cattle ranches and the Space Station, where
power generation on a small scale makes sense; but not by much.

What further scope is there for solar power? There are two main techniques for
trapping the sun's rays: by using them as a source of heat and by converting the
radiation into electricity. As with the solar constant the analysis is best expressed in
the unit familiar to all electricity consumers: the watt (W) and its metered equivalent the kilowatt hour (kWh). A 60 W bulb burning for 17 h uses just over 1 kWh
(60 x 17 = 1,020 W). Large units are usually expressed in megawatts or MW (that
is to say 1,000 kW) and the corresponding megawatt hours (MWh).

On an average day the Sun delivers the equivalent of about 1.4 kW per square
metre (m^2). At the top of the Earth's atmosphere. The quantity used to be called the
solar constant but, as we saw in Chapter 3, it varies during the 11-year cycle and,
though fairly steady during the year near the Tropics, it will fall close to zero during
winter at the poles and of course everywhere at night. About 8/10 of this wattage is
lost on its way to the ground because it is reflected by clouds or absorbed by gases

C. Vita-Finzi, *The Sun: A User's Manual*,
doi: 10.1007/978-1-4020-6881-2_8, © Springer Science+Business Media B.V. 2008

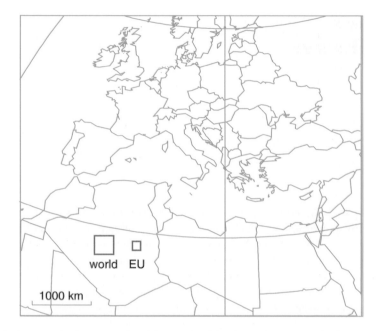

Fig. 8.1 Area required to meet electricity needs of the EU and the World in 2007 using solar thermal power plants (after www.desertec.org/fullneed.html)

in the atmosphere. Even so, the total solar energy that reaches the Earth's land areas in 1 day totals 2,000 times the energy consumed by humans during a whole year. The annual energy receipts for the Sahara are 2,500 kWh per m^2. In NW Europe the figure is nearer 900 kWh/m^2. As Isaac Asimov puts it, solar energy is copious but dilute (Fig. 8.1).

Passive solar power

Passive solar power refers to the use of the Sun's energy to heat or cool a building without any mechanical or electrical assistance. The key is to use the right materials, to site the building correctly and to plan the layout in order to promote or inhibit air circulation according to the needs of the moment.

Traditional building styles have often managed this with elegance and economy. In their *A Golden Thread – 2500 Years of Solar Architecture and Technology*, Ken Butti and John Perlin tell how Socrates observed that 'In houses that look toward the south, the sun penetrates the portico in winter' and the playwright Aeschylus remarked that only primitives and barbarians 'lacked knowledge of houses turned to face the winter sun, dwelling beneath the ground like swarming ants in sunless caves.' In Roman times the architect Vitruvius advised his readers to build winter rooms facing south, and laws were passed which banned the erection of buildings blocking such access to the winter sun. These principles are embodied in the adobe houses of the 12th century Pueblo Indians, notably at Acoma, or in the cliff dwellings of the Mesa Verde people in Colorado, which the low winter Sun but not the high Sun of summer can penetrate.

Growing numbers of modern buildings demonstrate how passive solar power can add flair to and subtract costs from tower blocks as well as family homes. Early practitioners include George Fred Keck and William Keck, who used air vents and wide eaves to ventilate and cool a number of private homes in the Chicago area in the 1930s, 1940s and 1950s. The first commercial office building with solar water heating and passive design is attributed to Frank Bridgers. The building, now known as the Bridgers-Paxton building and erected in Albuquerque, New Mexico, in 1956, was heated by a solar collector angled at 30° and facing S backed by insulated water heaters and heat pumps which could extract heat from stored water at temperatures down to 10°C. The building performed so well that, despite unusually cloudy conditions during the first year, only 8% of its heating requirements had to be supplied by the (electrically powered) heat exchanger.

Some of these approaches have been formalized by the US Department of Energy[2] and doubtless other government bodies. For instance, the benefits of passive solar heating are classed as direct gain, indirect gain and isolated gain. The first is illustrated by solar radiation which penetrates the living space and is stored in it. Indirect gain accumulates solar radiation using some kind of thermal storage device and then distributes it. Isolated gain systems collect and store heat in an area that can be opened to the rest of the house or closed off. Windows may consist of glass which provides the requisite solar heat gain coefficient (SHGC), conductive heat factor (U) and visible transmittance (VT). The SHGC measures the amount of solar radiation admitted directly or absorbed and subsequently released inward, and ranges from 0 to 1, so that low values represent shading ability.

In cold climates the SHGC should be 0.6 or more to maximize solar heat gain during winter, with U less than 0.35, ideally in combination with 'suntempering', the notion of orienting most of the building's glazing towards the south in the northern hemisphere and towards the north in the southern.

Shading, heat storage and natural cooling are other approaches which appear self evident but are often overlooked. They may be inhibited by street design or planning regulations but rarely completely ruled out. In Norman Foster's building at 30 St Mary Axe in the City of London, dating from 2004, mechanical air conditioning is required only 60% of the time thanks to the use of light wells for natural ventilation, floor-by-floor control of mechanical ventilation, and a double skin façade with blinds within the cavity. The light wells also reduce the need for artificial lighting (Fig. 8.2a–c). The cylindrical cross-section of the Burj al-Taqa (energy tower), a 68-storey building planned for Dubai by Eckhard Gerber, will reduce heat gain from the Sun; 60° of the surface will be covered by a solar shield and the entire skin will be made of a novel form of vacuum glass which transmits 2/3 the heat of current materials. Air cooled by sea water will be used for ventilation and, following the example of the wind towers in traditional Persian buildings, lateral openings will let in air which, being cooler, will sink into the building and displace warmer air within.

a

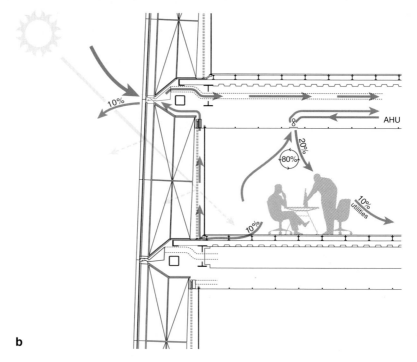

b

Fig. 8.2 (a) Swiss Re (30 St Mary Axe, London) in the London skyline (Centre; photograph by Nigel Young); (b) details of air conditioning in summer (After John Hewitt) (Both courtesy of Foster + Partners)

Solar heating

Of the active approaches to solar power the cheapest and most widespread at present is solar heating, where the Sun's rays raise the temperature of a liquid, usually water, which is then used directly or employed for space heating. Unconcentrated sunlight can heat fluids up to about 200°C, adequate for heating space and water. The first commercial solar heater for domestic use was patented in California in 1891. According to the California State government, by 1897 almost a third of the homes in Pasadena were equipped with solar water heaters; by 1920 they were being gradually replaced by heaters burning oil and natural gas. A similar cycle prevailed in Japan, where heavy reliance on solar power for domestic water heating was subverted, in the 1960s this time, by cheap oil imports, but successive oil crises in the 1970s encouraged a return to the Sun. In 2005 it was estimated that 10 m Japanese households used some form of solar heating.[3]

Domestic devices simply expose to the Sun a tank or pipe full of water which is at best painted black to improve its efficacy and fixed at the optimum angle for that particular location. Industrial versions may concentrate the radiation to heat oil or water which is then used to generate electricity or to distil water (Fig. 8.3). Concentration in parabolic troughs can yield temperatures of 400°C, parabolic dishes 650°C or even higher. Power towers where an array of mirrors reflects to a single receiver at the top of a tower can produce temperatures in excess of 1,500°C. An alternative design uses flat tracking mirrors to create an updraught which drives a set of turbines within the tower.

At Sanlúcar la Mayor, near Seville, in southern Spain (37.2° N) 624 movable mirrors or heliostats each with a surface area of 120 m² focus sunlight on water pipes on a tower 115 m high (Fig. 8.4). The resulting steam drives a turbine which

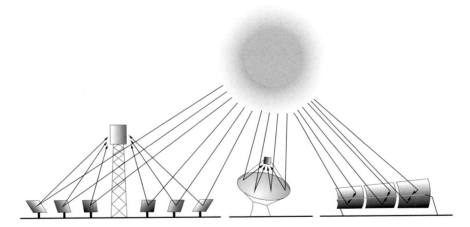

Fig. 8.3 (left to right) Concentrating the Sun using heliostats to focus radiation on a tower-mounted heat exchanger, a parabolic dish with a Stirling engine or other receiver at its focus, and parabolic troughs to heat a pipe containing oil which is then used to generate electricity or distil water

Fig. 8.4 11 MW Solar Platform at Sanlúcar la Mayor, Spain, first part of a 300 MW installation using a wide range of solar technologies, to be completed in 2013

generates 11 MW to a total of 20–25 GWh a year. Concentrating solar technology (CST) using steam had already been used at Adrano (Sicily), Almería (Spain) and Nio (Japan) in 1981, Barstow (California) and Targasonne (France) in 1982 and the Crimea in 1985. Sanlúcar clearly benefited from the experience of these early projects as well as experimental work in the USA, Israel and Spain, and opted for direct steam generation at 250°C and 40 bar (or atmospheres).

The mirrors track the Sun and they are curved to focus the radiation on a receiver at the top of the tower. As Spanish regulations do not permit more than 15% hybridisation of CST plants with fossil fuels, there is a limited 'thermal storage system' using molten salt which is kept topped up during normal operation and which allows the turbines to run for 50 min at 50% load during cloudy 'transient periods'.

PS10, the first of the plants at Sanlúcar, was built between 2001 and 2005. The intention is to bring total output in 2013 up to 300 MW, enough for 180,000 homes or the whole of Seville. The tower is already supplemented by a photovoltaic system (see below). The final complex is to include two additional thermoelectric towers each of 20 MW, five thermoelectric plants of 50 MW apiece of which some will be depend on towers and some on parabolic trough technology, and two further photovoltaic installations. The bulk of the funding for PS10 came from the European Commission and the rest from Central and Regional government sources.

The Mildura project in Australia, due to run at full capacity in 2013, is designed to use 50,000 mirrors over an area 6 by 7 km to concentrate sunlight by 500 times in PV modules which would then produce 1,500 times more power than equivalent roof-top cells. The cost is estimated to be close to that of a brown-coal power plant without the attendant environmental damage. The core of the system will be a tower 1 km high in which an updraught drives turbines near the base. The system is designed to have an output totalling 200 MW. The 'solar chimney' concept was tested at Manzanares, south of Madrid in Spain, between 1982 and 1989, with a tube 200 m high and 10 m across. The tower produced 50 kW; the efficiency of the system was 0.1% but the pilot study successfully demonstrated that the vortex embodied in the updraught was easy to control and displayed load swings which were less severe than in conventional windmills.[4]

The efficiency of solar heating is generally low. Although a solar dish driving a Stirling engine (which depends on heat differences in the system rather than any form of combustion) reportedly achieved over 40% efficiency in converting the heat to electricity, 14% is the accepted figure for solar radiation to net electric output where parabolic troughs are employed. We are of course only talking about running costs without taking into account the structure and the land on which it stands, staff, ground rent, insurance and other such items, but these are often omitted from estimates for other systems. But the technology is well tried and reasonably straightforward, even if some of its components, and in particular any tracking systems, are expensive to maintain. The way forward is evidently to start where solar energy is plentiful and demand is growing rapidly. Hence the large number of solar thermal power of projects reportedly under development in North Africa, southern Europe, Iran, Mexico and the USA, with a predicted global capacity of 36,000 MW by 2025 and a combined output of 100 million megawatt hours, equivalent to the total 2005 power consumption of Israel, Morocco, Algeria and Tunisia.[5]

Photovoltaics

The photovoltaic (PV) plant ancillary to the Sanlúcar tower, as we saw, also uses solar energy concentrated by tracking mirrors. Photovoltaic (PV) systems are more versatile but at present substantially more expensive to install than solar heating systems. They convert sunlight to electricity directly with the help of semiconductors, typically made of silicon which has been treated chemically to create a positive charge layer and a negative charge layer. When the PV cell is illuminated by sunlight electrons are liberated which give rise to an electric current. Devices for tracking the Sun are expensive.[6] But it also observed that solar energy could be useful for cooking in deserts, where firewood is in short supply, and that. That would seem the route to success: start where solar energy is plentiful and necessary. Among the first to exploit energy for domestic heat were Greece and Israel.

The PV effect was discovered by Alexandre-Edmond Becquerel in 1839 but it was not until the photovoltaic properties of silicon transistors were discovered at the Bell Laboratories in 1954 that the PV cell became a reality. The earliest applications were on orbiting satellites. It is said that solar cells were proposed in 1958 to back up battery power on the Earth-orbiting satellite Vanguard 1 and adopted against some opposition. They were still working 6 years later whereas the batteries lasted only 3 months.

Night is another matter. The full Moon delivers about 0.002 W/m², enough to upset the photoperiodic timing of some plants and to encourage others to take suitable evasive action. At the moment talk of exploiting this radiation is moonshine, even though the efficiency of PV systems is fast improving.

If (as with solar heating) we assess PV efficiency as the proportion of the incoming light energy that the cell renders available as electricity, the figure was 1% in 1883, when Charles Fritts used gold-plated selenium. The industry average in 2007 for single-crystalline silicon cells was 12–15%, for multicrystalline silicon 11–14% and for amorphous or thin-film silicon (where PV material is deposited on low-cost backing) 5–7%. Efficiency (in terms of voltage) falls with rising temperature by about 0.08 V per degree above 25°C, so that good ventilation is desirable. PV panels are also very sensitive to shading. If part of a panel is shaded total output is drastically reduced.

Efficiency can be increased to over 40% by the use of multi-junction cells, which consist of layers each of which captures the energy of a specified range of wavelengths, but any gain is greatly offset by the cost of production, which may be 100 times higher than for conventional PV materials. There are more effective semiconductor materials, such as gallium arsenide, which can attain efficiencies of over 25%, but their cost remains prohibitive for everyday use.[7]

Again as with solar heating, assessments of efficiency rarely take running costs into consideration. If all related expenditure in supplying and installing the panels is taken into account, the cost of 1 kW of solar electricity in early 2008 was 10 times dearer than the equivalent from a coal-fired plant. Payback – the time that elapses before the installation has paid for itself – is put by some commentators at 8 years in the UK. On the other hand the fuel comes gratis and maintenance is low.

A major drawback remains the high cost of purifying silicon, hence the popularity of thin-film cells. There is also a continuing search for alternative PV systems. In April 2007 a team at the University of New South Wales reported that a film of silver ~10 nm thick deposited on the solar cell and heated to 200°C broke into a series of 'islands' which improved light absorption and thus increased efficiency from 5–10% to 13–15%. Cells made from tiny silicon beads bonded in aluminium foil can be made into roofing tiles and promise to be half the price of crystalline cells. Even more audacious is the development in 1991 of dye-sensitized solar cells in 1991, with conversion efficiencies of about 11% and far lower costs of production than conventional PV materials. In 2007, experiments using organic dyes based on porphyrin, a constituent of chlorophyll and haemoglobin, indicate efficiencies of about 7%. The early work used dyes sensitive only to UV and blue frequencies; recent studies use dyes sensitive into the IR range.[8]

The price of traditional PV cells as a whole had fallen by 99.5% between 1958 and 1985; it fell by a further 80% between 1980 and 1988. Recent developments in California include printing solar cells on thin sheets of aluminium, with an estimated price in late 2007 of $1/W compared with $3.7/W for existing thin film modules. Global production of solar PV cells has risen sixfold since 2000, and by an additional 40% (2.3 GW) in 2007 alone. Although grid-connected solar capacity still provides less than 1% of the world's electricity, it increased by nearly 50% in 2006 to 5,000 MW thanks mainly to demand in Germany and Japan. Panels account for about half the total outlay. There is apparently little scope for rapid scaling up of panel production, and some of the dynamic newcomers, notably China, have still to convince potential clients that they would honour any liabilities arising from their systems. The first panels employing dye-sensitive material, the 'third-generation solar technology',[9] will come onto the market in 2010.

The lighter, new PV systems can be mounted on existing roofs or cheap foundations, and thin film versions can be wrapped around walls and towers. Architects and builders are beginning to use solar-cell material within roofs, façades, walls and windows in both new and existing buildings and to exploit rather than recoil from their colours and textures.[10] The lifetime of PV systems remains to be established but even now a minimum of 20 years is guaranteed by most manufacturers.

The benefits of a mass market and consumer demand that are illustrated by mobile telephones have yet to be seen in batteries for domestic PV installations or, indeed, in the cost of the panels themselves. Yet Germany shows what can be accomplished even in areas poor in sunshine. In 2007 its output – 2.5 GW – was 200 times that of the UK, thanks significantly by the feed-in tariff which guarantees payment of four times the market rate for 20 years for anyone who generates solar PV, wind or hydroelectric power.

In the UK, a set of PV panels with 10–15% efficiency could generate 150 kWh/year of DC electrical energy for every square metre of panel on a domestic roof, which, with the losses that arise in conversion to AC, represents 110 kWh/m^2. An average home uses about 3,300 kWh annually, so that 20 m^2 of south facing roof would suffice. Modern PV designs can be incorporated in complex and elegant roofs, as at the Eden project in Cornwall (Fig. 8.5). The Cooperative Insurance Society (CIS) service tower in Manchester, built in 1962, was initially covered with 14 million mosaic tiles. These began to fall off 6 months after the building was completed. In 2004–5 they were replaced by a weatherproof cladding (Fig. 8.6) which included 4898 80 W PV modules capable of generating 390 kW or 183,000 kWh a year.[11]

The potential applications are numerous even at present levels of efficiency and at current costs. Roof tiles have been developed which combine solar electric and solar thermal technologies. Small PV panels, already widely used for highway signs in the developed world, are beginning to show their value for ventilation systems in third world hospitals, earthquake monitoring systems in isolated locations, and rechargeable LED lamps for schools, clinics and water pumping. A fine illustration of what can be achieved is Sagar Island, in the SW of the Ganges delta, where since 1996 11 small PV plants provides electricity to a network of 14 villages 6–7 h every day.[12] But nationally, with a total installed PV capacity of 86 MW in mid 2006, India lags behind Japan, Germany and the USA.

Japan had reacted to the 'oil shocks' of 1973–4, when the Organization of Petroleum Exporting Countries (OPEC) raised the average export price of oil from $2.75 to $10 a barrel, by launching a 'sunshine project' to promote the development of PV systems. The current target is a national installation capacity of 4.82 GW

Fig. 8.5 The Core, Eden Project, Cornwall, UK. The solar energy roof has a Fibonacci Series design embodying 40 and 80 W PV panels and producing an estimated 20,000 kWh a year (Courtesy of solarcentury)

(i.e. 4,820 MW) by 2010 and the long-term ambition is to generate 50% of residential demand[13] with PV systems, equivalent to 100 GW by 2030. In 2004 China had a total PV installation of 60 MW (67% of its primary energy came from coal, and in 2006 two new major coal-fired plants were being added to the national grid each week). But things had begun to change in 2005, when a 'renewable energy promotion' law was passed committing the country to producing 15% of its power from clean energy sources by 2020. The present targets[14] for solar PV are 0.3 GW by 2010 and 30 GW by 2020. Perhaps more important, PV production in China tripled in 2006 and doubled in 2007 and it has displaced Germany in second place after Japan.[15] Moreover China easily leads the world in domestic solar hot water capacity[16] with 63% of the global total of 88 GW in 2007.

When it comes to large PV ground arrays Spain is supreme, with 2 MW systems, one in Murcia and one in Alicante, a 13.8 MW plant in Salamanca and a further large plant (12.7 MW) in Murcia. The USA fields a 14 MW installation at Nellis Air Force Base, Nevada, and is followed by Germany with a 12 MW system at Erlasee.

By January 2008 there were a further 850 or so large PV power plants in the world[17] with peak power output in excess of 170 kW. Their space requirements are not always easily met, although the tendency for arid areas to be thinly populated can be an advantage, especially if concentrating collectors combine solar tracking with the advantages of PV technology. The 11 MW solar power plant at Serpa, in SE Portugal, completed in January 2007, occupies 150 acres which are being rented from two farmers for

Fig. 8.6 The south façade of the CIS Tower in Manchester, UK, resurfaced in 2005 in PV panels which generate over 180,000 kWh a year (Courtesy of solarcentury)

25 years and accommodate modules with an efficiency of 12.6–17.7% which track the sun during the day. The power is fed directly to the national power grid. Funding came from three concerns, a Portuguese renewable energy company, a financial services organisation from the USA, which owns the plant, and a solar power system provider, also from the USA, which designed the plant and operates it. An attraction to investors was the Portuguese Government's scheme whereby a developer sells his power to utilities at a specified price.

To be sure, the need for large arrays to compensate for the inefficiency of PV devices creates problems as regards transmission as well as generation. In 1980 cost-effective transmission of electricity was limited to 7,000 km; with modern high voltage direct current transmission losses are estimated at about 3% per 1,000 km rising to perhaps 25% at 5,000 km. In compensation, bi-directional links are possible, so that peaks in solar output can be shared. An alternative is to generate hydrogen as fuel or as a component of fuel cells, as it can be piped or liquefied. In the Burj al-Taqa, excess electricity will generate hydrogen to be stored in tanks, and power at night will come from fuel cells. The passive devices on the Burj al-Taqa will be supplemented by wind turbines and by 17,000 m^2 of solar panels floating offshore.

Solar powered boats, cars and aircraft have been demonstrated. Perhaps the most striking yet sensible was *Solar Sailor*, a ferry powered by sun and wind, which operates in Sydney Harbour. The inspiration came to its designer, Robert Dane, who was watching 'a bunch of boffins' competing in the annual race for solar-powered boats on Lake Burley Griffin in Canberra. The clumsy devices he watched were hindered whenever the wind came up; Dane's boat uses its solar panels as sails whenever possible. The ferry has to deliver 10 knots for 10 h a day. At the moment it can manage 6 knots for 4 h, and it makes up the rest using a conventional engine. A similar boat is due to service the San Francisco-Alcatraz route from September 2008. It will save 30,000 l of diesel a year and emit no pollution when docking.

A modest solution

The biggest problem is inertia. In December 2007, for example, after many fine speeches, both Shell and BP relinquished their solar interests to concentrate on traditional concerns.[18] The inertia operates at all political and social levels and also at the design stage: the emphasis remains on further research when what is needed is the rapid manufacture of low-efficiency, economical PV material in bulk. In many parts of the world there is ample scope for the local production of roof-mounted solar heaters and turbines driven by fixed parabolic mirror systems.[18]

When discussing efficiency it pays to recall that the internal combustion engine rarely exceeds efficiencies of 30% and the steam engine typically attains 5–8%. Fortunately James Watt and Thomas Newcomen were more concerned with supplanting manual labour than with an unattainable ideal efficiency. Newcomen's 1712 beam engine had an efficiency of about 1% yet it was soon pumping water out of flooded mines; and comparison with internal combustion engines usually disregards the energy required to produce the fuel.

James Martin's acclaimed *The Meaning of the 21st Century*, published in 2006, devotes less than two pages out of a total 526 to solar power. It considers the fact that solar panels covering the area of the Nellis Air Force Base and its Nevada Test site could generate about twice the electricity needed by the entire USA – and then declares 'Nobody's suggesting this is a good idea.' If Martin is so dismissive, his many admirers, from James Lovelock to the *New Scientist*, must echo his preference for artificial clouds, sunshades in space, or the creation of a sulphuric acid aerosol in the stratosphere by the *addition* of sulphur to the fuel of airliners, as possible large-scale solutions to global warming.

Of course, cool, small countries are unlikely to embrace solar power as the primary, let alone the sole solution; the UK would have to sacrifice 12% of its territory to PV devices, Belgium 24%. The answer must lie in the trading of energies.

A cooperative scheme has been proposed for the European Union and the countries of the Middle East and North Africa, the TREC (Trans-Mediterranean Renewable Energy Cooperation) grid, which draws on solar, wind, geothermal, hydropower, biomass and conventional energy sources (Fig. 8.7). It foresees that by 2050, for instance, transmission lines with a capacity each of 5 GW will convey some 7×10^5 GWh a year of solar electricity from 20 regions in the Middle East and North Africa to the main centres of demand in Europe.[19] Sustainable energy will also be in heavy demand for water desalination in the countries best able to produce solar energy supplemented by wind. A sustainable energy world, with solar power as an important

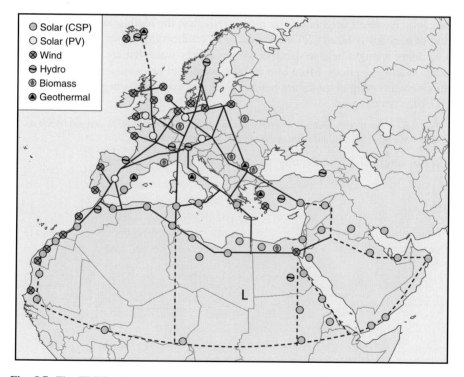

Fig. 8.7 The TREC scheme envisages a power transmission grid linking renewable energy sources in Europe, North Africa and the Middle East. A key component is the transfer of electricity to Europe from wind and concentrating solar power (CSP) installations in the deserts of North Africa and the Middle East along high tension DC wires (Courtesy desertec.org)

ingredient of the mix, is within reach provided solar storms or regional politics do not sabotage the enterprise.

Transmission of another kind could sidestep some of these problems. Several schemes are being proposed by which solar radiation is captured above the atmosphere and beamed directly down to receiving stations. The most ambitious is the proposal to base the power bases on the Moon.[20] In an alternative scheme, which builds on ideas first mooted in the 1960s and which has the support of the USA's National Security Space Office (NSSO) as well as commercial organisations, satellites in low Earth orbit would use microwaves to beam energy totalling somewhere between 10 and 25 MW, to a sequence of receiving stations on the ground.[21] A more modest version, which has the interests of energy-poor small island nations especially in mind, would exploit thin-film PV technology to transmit 1–1.2 MW to Earth; alternatively, a tethered satellite bearing a PV panel 2 by 1.9 km some 10 km above the Earth's surface[22] could reportedly deliver one GW.

The reaction of many commentators to such proposals is predictably cool. Why, they ask would Europe put power plants in space when there is so much sun in the Mediterranean countries and in the North African desert?'[23] To which one might add 'why delay taking action when the technology for harnessing solar energy both locally and regionally is already here?' Consider the case of Libya, which could share with its desert neighbours, far more needy than the Mediterranean nations, the fruits of a modestly-sized CST array sited in its southeastern corner (L on Fig. 8.7).[23] In so doing it might help to damp down the energy wars that are being fuelled by rising oil and gas prices. More important, it would set an example to other oil-rich nations by investing in the future before its oil and water wells begin to gasp.

Galileo once remarked.

The Sun, with all those planets revolving around it and dependent on it, can still ripen a bunch of grapes as if it had nothing else in the universe to do.

It can surely also find the time to sort out our energy problems.

Endnotes

Chapter 1

1. Mellars 2004
2. Tudge 2005
3. Ray 1989
4. Ghezzi and Ruggles 2007
5. Lockyer 1909
6. Freeth et al. 2006
7. Renfrew and Bahn 1991; Marshack 1991
8. Boorstin 1983
9. Hall 1967
10. Vita-Finzi 2002
11. Hawkes 1962
12. de la Vega 1959
13. B. van der Spek 2007
14. Lendering 2007
15. Needham 1959
16. van Helden 1995
17. Jones 2001
18. van Helden 1995
19. Bray and Loughhead 1964
20. Asimov 1987
21. Zirin 1992
22. Hufbauer 1991
23. Hawkes 1962
24. Bhatnagar and Livingston 2005
25. Papastefanou and Ioannidou 2004

Chapter 2

1. *Pliny, Natural History,* bkii, ch 59; Diogenes Laertius 272
2. Hippolytus 2004, 1, epitome Ch 7
3. Jeffreys 1970
4. Miller 2005
5. Dyson et al. 1920

6. Parkinson 1996
7. Alpher et al. 1948
8. Bhatnagar and Livingston 2005; Haubold and Mathai 1997
9. Newkirk 1980
10. Bahcall 2000
11. Demarque and Guenther 1999
12. Bahcall 2000
13. Parker 2000
14. Kaufmann and Freedman 1999
15. Noyes 1990
16. Lean 2005
17. Parker 2000
18. Pogge 2003
19. Dick 1982
20. Hufbauer 1991
21. Herschel 1795
22. Sagan and Mullen 1972; Holland 1984
23. McMillan 1997
24. Mayor and Queloz 1995
25. McMillan 1997
26. Tinetti et al. 2007
27. Willman 2007

Chapter 3

1. see discussion in www.earthsci.unimelb.edu.au
2. Gaidos et al. 2000
3. Schonberg and Chandrasekhar 1942
4. www.ESA.int, March 2007
5. Radick 2004
6. Caffee et al. 1985
7. Mackay 2008
8. Crozaz et al. 1977
9. Ouyang Ziyuan in *China Daily* 26/07/2006; Vita-Finzi 2008
10. Vita-Finzi 2008
11. Webber and Higbie 2003
12. Bray and Loughhead 1964
13. Shirley and Fairbridge 1997
14. Bray and Loughhead 1964
15. Foukal 1990
16. Lang 2006
17. Cullen 1980
18. Beckman and Mahoney 1998
19. Radick 2004
20. Eddy 1983
21. Lean 1991
22. Stuiver et al. 1998
23. Suess 1980; Stuiver and Becker 1993; Dergachev 2004
24. de Jaeger 2005.
25. Willson and Mordvinov 2003

26. Maunder 1890
27. Davis 1981
28. Eddy 1977
29. Silverman 1992
30. Carrington 1859
31. Reames 1995
32. Leighton et al. 1961
33. Gilliland 1982
34. Ribes et al. 1987
35. Lockwood et al. 1999
36. Gingerich 1989

Chapter 4

1. Parker 1999
2. McCormack and North 2004
3. In a burst of aptness a student once stated in an examination that (the planet) Venus has a high libido
4. Harper 2004
5. Kwa 2000
6. Pavlov et al. 2000
7. Sackmann and Boothroyd 2003
8. Priem 1997
9. Croll 1875
10. Zeuner 1959
11. e.g. Lockwood et al. 2007
12. Williams 1990
13. Kirkby 2002
14. Drake and Bristow 2006
15. Liu et al. 2007; Claussen et al. 1999; deMenocal et al. 2000
16. Grove and Switsur 1994
17. Burckle and Grissino-Mayer 2003
18. Houghton et al. 2001
19. Friis-Christensen and Lassen 1991
20. Damon and Laut 2004
21. Lockwood and Fröhlich 2007
22. Reid 1997
23. Kirkby 2002
24. The lead scientist is Jasper Kirkby. Report by Lawrence Solomon *Financial Post* 23 February 2007
25. Houghton 1986
26. Fligge and Solanki 2000
27. Haigh 1996
28. Lamb 1995
29. Tinsley 1988
30. Herman and Goldberg 1978
31. Vita-Finzi 2008
32. Grove and Rackham 2001
33. Haynes 1968
34. Force 2004

35. Leopold and Vita-Finzi 1998
36. Wallén 1955

Chapter 5

1. E g www-spof gsfc nasa gov/stargaze/Sun1lie htm 2006
2. Brack 2007
3. Hart-Davis 2004
4. Lotka 1956
5. Benner et al. 2004
6. van Dover 1996
7. Swartz et al. 2007
8. Benner et al. 2004
9. Darwin 1871
10. Miller 1953
11. C. Sagan quoted by Shapiro 1986, p. 99
12. Tsujimoto et al. 2002
13. Cockell 2002
14. Cnossen et al. 2007 especially Fig. 7
15. Rothschild 1999; 2003
16. Sagan 1973; Cockell 1998
17. Peterson and Côté 2004
18. Rothschild 1999; 2003
19. Vázquez and Hanslmeier 2006
20. Schopf 1999
21. Wolstencroft and Raven 2002
22. Harrison 1973
23. Lucas et al. 2005
24. Encyclopedia Britannica 1972
25. Brack 2007
26. Burchell MJ et al. 2001
27. R. Mancinelli www. space. com
28. Horneck et al. 1994
29. Zahradka et al. 2006
30. Adrian Melott, cited by University of Kansas News Release, 18 May 2007
31. Medvedev and Melott 2007; Gies and Helsel 2005
32. Hamilton and Buchanan 2007
33. Leopold and Vita-Finzi 2005
34. Dillehay 1984
35. Martin 1973
36. Brown et al. 2004
37. Meiri 2007
38. Culotta 2007; Lalueza-Fox 2007
39. Harrison et al. 1977; Jablonski and Chaplin 2000
40. Wheeler 1994
41. Clarke 1993
42. Arrese et al. 2006
43. Huntington 1945
44. deMenocal 2001
45. Lyell 1830, I, 164-165

Chapter 6

1. Zhong 2007
2. Vallina and Simó 2007
3. Lean 1997
4. UV radiation guide 1992 www-nehc.med.navy.mil/downloads/ih/uvdoc2.doc
5. WHO 1994
6. Lovelock 1972
7. Farman et al. 1985
8. Harm 1980
9. Singer 1993
10. Herman et al. 2000
11. Rosenthal et al. 1984
12. Sher 2001
13. Danilenko et al. 1994
14. Lam 1998
15. Glickman et al. 2006
16. Gambichler et al. 2002
17. Wharton and Cockerell 1998
18. Keystone et al. 2004
19. Kudish et al. 1997
20. Randle 1997
21. Porter 1997
22. Randle 1997
23. Eide and Weinstock 2005
24. Eide and Weinstock 2005
25. Scheinfeld and DeLeo 2005
26. Naylor and Rigel 2005
27. Eide and Weinstock 2005
28. Graedel and Crutzen 1993
29. Gillie 2004
30. Porter 1997
31. Brit J Nursing 1935, 83: 311
32. Robinson et al. 2005
33. Engelsen and Webb 2007
34. Gillie 2004
35. Ball 2004
36. Grant et al. 2008
37. Smith 1992
38. John Aryankalayil et al. 2006
39. Taylor 1988
40. Mukesh et al. 2006
41. WHO 1994
42. WHO 2001

Chapter 7

1. NASA 2007
2. Schlegel 2006
3. Cliver 1994

4. In Cliver 1994
5. Bell 1880
6. Kharkov 2007; Lanzerotti 2001
7. Angot 1897
8. Canada 2006
9. Marconi 1928 in Lanzerotti 2001
10. Kharkov 2007
11. *The Guardian* 16 Nov 2007
12. Sharifi and Arozullah 1986
13. Lanzerotti 2001
14. Lanzerotti 2001
15. Wright 2007
16. Gorney 1990
17. Ken Phillips personal communication 22 Nov 2007
18. Vita-Finzi 2002
19. Selleri 1998
20. Foulger and Hofton 1998
21. A. Cerruti and P. Kintner reported in *New Scientist* 29 Sept 2006
22. Ken Phillips, personal communication 22 Nov 2007
23. Boerner et al. 1983
24. Kappenman et al. 1997
25. J. Brooks, The Subtle Storm *New Yorker* 7 Feb 1959, cited by Lanzerotti 2001
26. Wikipedia 2008
27. Hydro-Québec 2007
28. Marusek 2007
29. Boerner et al. 1983
30. Campbell 1978
31. Hydro-Québec 2007
32. *Le terrible aveu des hommes de la lune à leurs femmes*
33. Baglioni et al. 2007
34. www solarstorms org
35. Lucid 1998
36. NASA 2004
37. National Research Council 2006
38. Marusek 2007
39. Cucinotta 2004
40. Churchley and Sung 2006
41. Hathaway et al. 1999
42. Ogurtsov 2005
43. D Baker, Laboratory for Atmospheric and Space Physics, University of Colorado, Boulder, reported on 25 April 2007 by CNN.com
44. Jones et al. 2005
45. Posner 2007
46. At www.ips.gov.au/Space_Weather

Chapter 8

1. Crabtree & Lewis 2007
2. US Dept of Energy 2000
3. Perlin 2005
4. Michaud 1999

  5. Greenpeace 2005
  6. van Dulken 2002
  7. US Department of Energy 2006
  8. Dye sensitized solar cells (DYSC) based on nanocrystalline oxide semiconductor film (lpi. epfl.ch/solarcellE.html)
  9. www.dyesol.com
 10. See for example Hermansdörfer and Rüb 2005
 11. www.solarcentury.co.uk
 12. West Bengal Renewable Energy Development Agency (www.wbreada.org)
 13. EPIA/Greenpeace 2004
 14. www.martinot.info/china.htm#targets
 15. Dorn 2007
 16. New Scientist 10 Nov 2007
 17. Lenardic 2008
 18. The Guardian 11 Dec 2007
 19. Trieb and Müller-Steinhagen 2007
 20. Sasaki et al. 2004
 21. Space Island Group (www.spaceislandgroup.com)
 22. epi.epfl.ch/solarcellE.html
 23. Summerer 2006

References

Alpher RA et al. (1948) The origin of chemical elements. Phys Rev 73:803–804

Angot J (1897) The Aurora Borealis. Appleton, New York

Arrese C et al. (2006) Behavioural evidence for marsupial trichromacy. Curr Biol 16: R193–R194

Asimov I (1987) Asimov's New Guide to Science. Penguin, Harmondsworth

Baglioni P , Sabbatini M, Horneck G (2007) Astrobiology experiments in low earth orbit – facilities, instrumentation and results. In: Horneck G, Rettberg P (eds.) Complete Course in Astrobiology. Wiley, Weinheim, pp 273–320

Bahcall JN (2000) How the Sun shines. nobelprizeorg/nobel_prizes/physics/articles/fusion/index html

Ball GFM (2004) Vitamins. Blackwell, London

Beckman JE, Mahoney TJ (1998) The Maunder Minimum and climatic change. www stsci edu/ stsci/meetings

Bell AG (1880) On the production and reproduction of sound by light. Am J Sci 20:305–324

Benner S et al. (2004) Is there a common chemical model for life in the universe? Curr Opin Chem Biol 8:672–689

Bhatnagar A, Livingston W (2005) Fundamentals of Solar Astronomy. World Scientific, Hackensack, NJ

Boerner WM et al. (1983) Impacts of solar and auroral storms on power line systems. Space Sci Rev 35:195–205

Boorstin DJ (1983) The Discoverers. Random House, New York

Brack A (2007) Astrobiology: from the origin of life on Earth to life in the Universe. In: Horneck G, Rettberg P (eds) Complete Course in Astrobiology. Wiley, Weinheim, pp 1–22

Bray RJ, Loughhead RE (1964) Sunspots. Constable, London

Brown P et al. (2004) A new small-bodied hominin from the Late Pleistocene of Flores, Indonesia. Nature 431:1055–1061

Burchell MJ et al. (2001) Survivability of bacteria in hypervelocity impact. Icarus 154:545–547

Burckle L, Grissino-Mayer HD (2003) Stradivari, violins, tree rings, and the Maunder Minimum: a hypothesis. Dendrochron 21:41–45

Butti K, Perlin J (1980) A golden thread: 2500 years of solar architecture and technology. Van Nostrand Reinhold, New York

Caffee MW et al. (1985) Evidence in meteorites for an active early Sun. NASA Tech Rep 19850018239

Campbell WH (1978) Induction of auroral zone electric currents within the Alaska pipeline. Pure Appl Geophys 116:1143–1173

Canada (2006) 150 years of geomagnetic effects www spaceweather gc ca/historyeffects_e php

Carrington R (1859) Description of a singular appearance seen in the Sun on September 1, 1859. Mon Not Roy Astronom Soc 20:13–15

Churchley L, Sung M (2006) Cosmic radiation and flying. WHO Information Sheet, Med Phys 777

Clarke AC (1993) By Space Possessed. Gollancz, London

Claussen M et al. (1999) Synergistic feedbacks from ocean and vegetation on the African monsoon response to mid-Holocene insolation. Geophys Res Lett 26:2481–2484

Cliver EW (1994) Solar activity and geomagnetic storms: the first 40 years. EOS 75:569, 574–575

Cnossen I et al. (2007) Habitat of early life: Solar X-ray and UV radiation at Earth's surface 4–3 5 billion years ago. J Geophys Res 112, E02008, doi: 101029/2006JE002784

Cockell CS (1998) Biological effects of high ultraviolet radiation on early Earth: a theoretical evaluation. J Theor Biol 193:717–729

Cockell CS (2002) Photobiological uncertainties in the Archaean and post-Archaean world. Int J Astrobiol 1:31–38

Crabtree GW, Lewis NS (2007) Solar energy conversion. Phys Today 60:37–42

Croll J (1875) Climate and Time in their Geological Relations. Daldy, Tbister & Co., New York

Crozaz G et al. (1977) The record of solar and galactic radiations in the ancient lunar regolith and their implications for the early history of the Sun and Moon. Phil Trans Roy Soc A 285:587–592

Cucinotta FA (2004) Can people go to Mars? http://science.nasa.gov/headlines/y2004/17feb_radiation.htm

Cullen C (1980) Was there a Maunder Minimum? Nature 283:427–428

Culotta E (2007) Ancient DNA reveals Neanderthals with red hair, fair complexions. Science 318: 546–547

Damon PE, Laut P (2004) Pattern of strange errors plagues solar and terrestrial climate data. EOS 85, 370, 374

Danilenko KV et al. (1994) Diurnal and seasonal variations of melatonin and serotonin in women with seasonal affective disorder. Arctic Med Res 53:137–145

Darwin CR (1871) Letter to JD Hooker 1 February 1871

Davis TN (1981) Red Aurora 19 December 1980. www.gi.alaska.edu/scienceforum/ASF4

de Jaeger C (2005) Solar forcing of climate. 1: solar variability. Space Sci Rev 120:197–241

de la Vega Garcilaso Inca (1959) Comentarios Reales de los Incas, Estudio preliminary notas de José Durand (trans. James Q. Jacobs). Univ Nac Mayor de San Marcos, Lima, Peru

Demarque P, Guenther DB (1999) Helioseismology: probing the interior of a star. Proc Natl Acad Sci 96:5356–5359

deMenocal P (2001) Cultural responses to climate change during the late Holocene. Science 292:667–673

deMenocal P et al. (2000) Coherent high- and low-latitude climate variability during the Holocene warm period. Science 288:2198–2202

Dergachev VA (2004) Manifestation of the long-term solar cyclicity in climate archives over 10 millennia. Proc IAU Symp 223:699–704

Dick SJ (1982) Plurality of Worlds. Cambridge University Press, Cambridge

Dillehay T (1984) A late Ice-Age settlement in southern Chile. Sci Am 251:100–109

Dorn J (2007) Solar cell production jumps 50 percent in 2007. www.earth-policy.org/solar/2007.htm.

Drake N, Bristow C (2006) Shorelines in the Sahara: geomorphological evidence for an enhanced monsoon from palaeolake Megachad. The Holocene 16:901–911

Dyson FW et al. (1920) A determination of the deflection of light by the Sun's gravitational field, from observations made at the total eclipse of May 29, 1919. Phil Trans Roy Soc A 220:291–333

Eddy JA (1977) The case of the missing sunspots. Sci Am 236:80–92

Eddy JA (1983) The Maunder Minimum; a reappraisal. Solar Phys 89:195–207

Eide MJ, Weistock MA (2005) Epidemiology of skin cancer. In: Rigel DS et al. (eds) Cancer of the Skin. Elsevier Saunders, Philadelphia, PA, pp 47–49

Encyclopedia Britannica (1972) Entry for Stereochemistry. Benton, Chicago, IL

Engelsen O, Webb AR (2007) Review: the relationship between UV exposure and vitamin D status. Proc UV Conf 2007, Davos, Switzerland, pp 123–124

EPIA/Greenpeace (2004) Solar Generation. Greenpeace, Brussels

Farman J et al. (1985) Large losses of total ozone in Antarctica reveal seasonal C10x/NOx interaction. Nature 315:207–210

Fligge M, Solanki SK (2000) Modelling short-term spectral irradiance variations. Space Sci Rev 94:139–144

Force ER (2004) Late Holocene behavior of Chaco and McElmo canyon drainages (Southwest U.S.): a comparison based on archaeologic age controls. Geoarch 19:583–609

Foukal P (1990) Solar Astrophysics. Wiley, New York

Foulger GR, Hofton MA (1998) Regional vertical motion in Iceland 1987–1992, determined using GPS surveying. In: Stewart IS, Vita-Finzi C (eds) Coastal Tectonics. Geological Society of London Special Publications 146, pp 165–178

Freeth T et al. (2006) Decoding the ancient Greek astronomical calculator known as the Antikythera Mechanism. Nature 444:587–591

Friis-Christensen E, Lassen K (1991) Length of the solar cycle: an indicator of solar activity closely associated with climate. Science 254:698–700

Gaidos EJ et al. (2000) The Faint Young Sun Paradox: an observational test of an alternative solar model. Geophys Res Lett 27:501–504

Gambichler T et al. (2002) Plasma levels of opioid peptides after sunbed exposures. J Dermatol 147:1207–1211

Ghezzi I, Ruggles C (2007) Chankillo: a 2300-year-old solar observatory in coastal Peru. Science 315:1239–1243

Gies DR, Helsel JW (2005) Ice Age epochs and the Sun's path through the Galaxy. Astrophys J 626:844–848

Gillie O (2004) Sunlight Robbery. Health Research Forum, London

Gilliland RL (1982) Solar, volcanic and CO_2 forcing of recent climatic change. Clim Change 4:111–131

Gingerich O (1989) Johannes Kepler. In: Taton R, Wilson C (eds) General History of Astronomy 2A. Cambridge University Press, Cambridge, pp 54–78

Glickman G et al. (2006) Light therapy for seasonal affective disorder with blue narrow-band light-emitting diodes (LEDs). Biol Psych 59:502–507

Gorney DJ (1990) Solar cycle effects on the near-Earth space environment. Rev Geophys 28:315–336

Graedel TE, Crutzen PJ (1993) Atmospheric Change. Freeman, New York

Grant WB et al. (2008) Comparisons of estimated economic burdens due to insufficient solar ultraviolet irradiance and vitamin D and excess solar UV irradiance for the United States. Photochem Photobiol 81:1276–1286

Greenpeace (2005) Concentrated solar thermal power – now! Greenpeace International, Amsterdam

Grove AT, Rackham O (2001) The Nature of Mediterranean Europe. Yale University Press, New Haven, CT

Grove JM, Switsur R (1994) Glacial geological evidence for the medieval warm period. Clim Change 30:1–27

Haigh J (1996) The impact of solar variability on climate. Science 272:981–984

Hall DN (1967) Further notes on navigating in deserts. Geog J 133:508–511

Harm W (1980) Biological effects of ultraviolet radiation. Cambridge University Press, Cambridge

Harper KC (2004) The Scandinavian tag-team: providers of atmospheric reality to numerical weather prediction efforts in the United States (1948–1955). Proc Int Comm Hist Met 1:84–90

Harrison LG (1973) Evolution of biochemical systems with specific chiralities: a model involving territorial behaviour. J Theor Biol 39:333–341

Harrison GA et al. (1977) Human Biology. Oxford University Press, Oxford

Hart-Davis A (ed) (2004) Talking Science. Wiley, Chichester

Hathaway DH et al. (1999) A synthesis of solar cycle prediction techniques. J Geophys Res 104:22,375–22,388

Haubold HJ, Mathai AM (1997) The Sun. In: Shirley JH, Fairbridge RW (eds) Encyclopedia of Planetary Science, Kluwer, Dordrecht, pp 786–794

Hawkes J (1962) Man and the Sun. Cresset, London

Haynes CV Jr (1968) Geochronology of Late-Quaternary alluvium. In: Morrison RB, Wright HE (eds) Means of Correlation of Quaternary Successions. University of Utah Press, Salt Lake City, UT, pp 591–631

Herman JR, Goldberg R (1978) Sun Weather and Climate. Dover, New York

Herman J et al. (2000) Ultraviolet exposure (UV-B) perturbations caused by the quasi-biennial oscillations (QBO) in ozone. J Geophys Res 105:29189–29193

Hermansdörfer I, Rüb C (2005) Solar Design. Jovis Verlag, Berlin

Herschel W (1795) On the nature and construction of the sun and fixed stars. Phil Trans Roy Soc Lond 85:46–72

Hippolytus (2004) The Refutation of all Heresies. Kessinger, Whitefish, MT

Holland HD (1984) The Chemical Evolution of the Atmosphere and Ocean. Wiley, New York

Horneck G et al. (2004) Long-term survival of bacterial spores in space. Adv Space Sci 14:41–45

Houghton JT (1986) The Physics of Atmospheres (2nd ed). Cambridge University Press, Cambridge

Houghton JT et al. (eds) (2001) Climate Change 2001: The Scientific Basis Cambridge University Press, Cambridge

Hufbauer K (1991) Exploring the Sun. Johns Hopkins University Press, Baltimore, MD

Huntington E (1945) Mainsprings of Civilization. Wiley, New York

Hydro-Québec (2007) www.hydroquebec.com

Jablonski NG, Chaplin G (2000) The evolution of human skin coloration. J Hum Evol 39:57–106

Jeffreys H (1970) The Earth (5th ed) Cambridge University Press, Cambridge

John-Aryankalayil M et al. (2006) Microarray and protein analysis of human pterygium. Mol Vision 12:55–64

Jones N (2001) Spot on. New Sci 169, 28 July, 12

Jones JBL et al. (2005) Space weather and commercial airlines. Adv Space Sci 36:2258–2267

Kappenman JG et al. (1997) Geomagnetic storms can threaten electric power grids. Earth Space 9:9–11

Kaufmann WJ III Freedman RA (1999) Universe (5th ed). Freeman, New York

Keystone JS et al. (eds) (2004) Travel Medicine. Mosby, Edinburgh

Kharkov (2007) Solar Activity and Space Weather. Kharkov Astronomical Observatory, Ukraine

Kirkby J (2002) Cloud: a particle beam facility to investigate the influence of cosmic rays on clouds. Proc IACI Workshop. CERN 2001–2007

Kudish AO (1997) The analysis of ultraviolet radiation in the Dead Sea basin, Israel. Int J Clim 17:1697–1704

Kwa C (2000) The rise and fall of weather modification. In: Miller CA, Edwards PN (eds) Changing the Atmosphere: Expert Knowledge and Environmental Governance. MIT, Cambridge, MA, pp 135–165

Lalueza-Fox C et al. (2007) A melacortin 1 receptor allele suggests varying pigmentation among Neanderthals. Science 318:1453–1455

Lam RW (ed) (1998) Seasonal Affective Disorder and Beyond. Am Psych Press, Washington, DC

Lamb HH (1995) Climate, History and the Modern World. Routledge, London

Lang KR (2006) Sun, Earth and Sky (2nd ed). Springer, New York

Lanzerotti LJ (2001) Space weather effect on technologies. In: Song P et al. (eds) Space Weather. Geophys Monog 125, American Geophysical Union, Washington, DC, pp 11–22

Lean J (1991) Variations in the Sun's radiative output. Rev Geophys 29:505–535

Lean J (1997) The Sun's variable radiation and its relevance for Earth. Ann Rev Astro Astrophys 35:33–67

Lean J (2005) Living with a variable Sun. Phys Today 58:32–38

Leighton R et al. (1961) Velocity fields in the solar atmosphere. Astrophys J 135:474–499

Lenardic D (2008) Photovoltaic systems-technologies and applications. http://www.pvresources. com/en/top50pv.php).

Lendering J (2007) Alexander's final days: a Babylonian perspective. www livius org

Leopold LB, Vita-Finzi C (1998) Valley changes in the Mediterranean and America and their effects on humans. Proc Am Phil Soc 142:1–17

Leopold LB, Vita-Finzi C (2005) Archeological trash. Catena 62:1–13

Liu Z et al. (2007) Simulating the transient evolution and abrupt change of Northern Africa atmosphere-ocean-terrestrial ecosystem in the Holocene. Quat Sci Rev 26:818–1837

Lockwood GW et al. (2007) Patterns of photometric and chromospheric variation among Sun-like stars: a 20 year perspective. Astrophys J, Suppl Ser 171:260–303

Lockwood M, Fröhlich C (2007) Recent oppositely directed trends in solar climate forcings and the global mean surface air temperature. Proc Roy Soc A 463, 2447–2460

Lockwood M et al. (1999) A doubling of the Sun's coronal magnetic field during the last 100 Years. Nature 399:437–439

Lockyer N (1909) Stonehenge and Other British Stone Monuments Astronomically Considered. Macmillan, London

Lotka AJ (1956) Elements of Mathematical Biology (1st pub 1924). Dover, New York

Lovelock J (1972) Gaia as seen through the atmosphere. Atmos Env 6:579–580

Lucas PW et al. (2005) UV circular polarisation in star formation regions: the origin of homochirality? Orig Life Evol Bios 35:29–60

Lucid SW (1998) Six months on Mir. Sci Am 278:46–57

Lyell C (1830) Principles of Geology, I. Murray, London

McCormack G, North GR (2004) Solar variability and climate. In: Pap J, Fox P (eds) Solar Variability and Its Effects on Climate. Geophys Monog 141, American Geophysical Union, Washington, DC

McKay DS (2008) The Moon as a giant tape recorder for solar system and solar events. www.lpi. usra.edu/meetings/LEA/whitepapers/McKay.pdf

McMillan RS (1997) Planet: extrasolar. In: Shirley JH, Fairbridge RW (eds) Encyclopedia of Planetary Science. Kluwer, Dordrecht, pp 588–589

Martin PS (1973) The discovery of America. Science 179:969–974.

Marshack A (1991) The Taï plaque and calendrical notation in the Upper Paleolithic. Camb Arch J. 1:25–61

Marusek JA (2007) Forecasting solar storms in Solar Cycles 24 & 25. http://personals.galaxy internet.net/tunga/SSTA.pdf

Maunder EW (1890) Professor Spoerer's researches on Sun-spots. Mon Not Roy Astr Soc 50: 251–252

Mayor M, Queloz D (1995) A Jupiter-mass companion to a solar-type star. Nature 378:355

Medvedev MV, Melott A (2007) Do extragalactic cosmic rays induce cycles in fossil diversity? Astrophys J 664:879–889

Meiri S et al. (2007) The island rule: made to be broken?. Proc Roy Soc B 275:141–148

Mellars P (2004) Reindeer specialization in the early Upper Palaeolithic: the evidence from south west France. J Arch Sci 31:613–617

Michaud LM (1999) Vortex process for capturing the mechanical energy produced during upward heat convection in the atmosphere. Appl Energy 62:241–251.

Miller SL (1953) A production of amino acids under possible primitive Earth conditions. Science 117:528–529

Miller AI (2005) Empire of Stars. Little, Brown, London

Mukesh BN et al. (2006). Development of cataract and associated risk factors: the Visual Impairment Project. Arch Ophthalmol 124:79–85

NASA (2004) science.nasa.gov/headlines/y2004/17feb_radiation.htm

NASA (2007) sohowww nascom nasa gov/spaceweather

National Research Council (2006) Space Radiation Hazards and the Vision for Space Exploration: Report of a Workshop, Ad Hoc Committee on the Solar System Radiation Environment and NASA's Vision for Space Exploration

Naylor MF, Rigel DS (2005) Current concepts in sunscreens and usage. In: Rigel DS et al. (eds) Cancer of the Skin. Elsevier Saunders, Philadelphia, PA, pp 71–80

Needham J (1959) Science and Civilisation in China, v 3. Cambridge University Press, Cambridge

Newkirk G Jr (1980) Solar variability on time scales of 10^5 years to $10^{9.6}$ years. In: Pepin RO et al. (eds) Proc Conf Ancient Sun, Geochim Cosmochim Acta Suppl 13, pp 293–320

Noyes RW (1990) The Sun. In: Beatty JK et al. (eds) The New Solar System. Cambridge University Press, Cambridge, pp 15–28

Ogurtsov MG (2005) On the possibility of forecasting the Sun's activity using radiocarbon solar proxy. Solar Phys 231:167–176

Papastefanou C, Ioannidou A (2004) Beryllium-7 and solar activity. Appl Rad Isotopes 61:1493–1495

Parker EN (1999) Sunny side of global warming. Nature 399:16–18

Parker EN (2000) The physics of the Sun and the gateway to the Stars. Phys Today 53:26–31

Parkinson BW (1996) Global Positioning System. American Institute of Aeronautics and Astronautics, Washington, DC

Pavlov AA et al.(2000) Greenhouse warming by CH_4 in the atmosphere of early Earth. J Geophys Res - Planets 105:11,981–11,990

Perlin J (2005) Solar evolution. www.solarschoolhouse org/history html

Peterson CL, Côté J (2004) Cellular machineries for chromosomal DNA repair. Genes Dev 18:602–616

Pogge RW (2003) The folly of Giordano Bruno. Guest editorial www.setileague org

Porter R (1997) The Greatest Benefit to Mankind. HarperCollins, London

Posner A (2007) Up to 1-hour forecasting of radiation hazards from solar energetic ion events with relativistic electrons. Space Weather 5, doi:10 1029/2006SW000268

Priem HNA (1997) CO_2 and climate: a geologist's view. Space Sci Rev 81: 173–198

Radick RR. (2004) Long-term solar variability: evolutionary time scales In: Pap J, Fox P (eds) Solar Variability and its Effects on Climate, Geophys Monog 141, Am Geophys Un, Washington, DC, pp 9–14

Randle HW (1997) Suntanning: differences in perceptions throughout history. Proc Mayo Clin 72:461–466

Ray TP (1989) The winter solstice phenomenon at Newgrange, Ireland: accident or design? Nature 337:43–345

Reames DV (1995) Solar energetic particles: a paradigm shift. Rev Geophys (Suppl) 33:585–590

Reid G (1997) Solar forcing of global climate change since the mid-17th century. Clim Change 37:391–405.

Renfrew C, Bahn PG (1991) Archaeology. Thames & Hudson, London

Ribes E et al. (1987) Evidence for larger Sun with a slower rotation during the seventeenth century. Nature 326:52–55

Robinson PD et al. (2005) The re-emerging burden of rickets: a decade of experience from Sydney. Arch Dis Child 91:564–568

Rosenthal E et al.(1984) Seasonal Affective Disorder: a description of the syndrome and preliminary findings with light therapy. Arch Gen Psych 41:72–80

Rothschild LJ (1999) The influence of UV radiation on protistan evolution. J Eukar Microbiol 46:548–555

Rothschild LJ (2003) The sun: the impetus of life. In: Rothschild LJ, Lister A (eds) Evolution on Planet Earth, Academic, London, pp 87–107

Sackmann IJ, Boothroyd AI (2003) Our Sun. V. A bright young Sun consistent with helioseismology and warm temperatures on ancient Earth and Mars. Astrophys J 583:1024–1039

Sagan C (1973) Ultraviolet selection pressure on the earliest organisms. J Theor Biol 39:195–200.

Sagan C, Mullen G (1972) Earth and Mars: evolution of atmospheres and surface temperatures. Science 177:52–56

Sasaki S et al. (2004) Engineering research for tethered solar power satellite. Proc Radio Sci Conf 2004, 607–615

Scheinfeld N, DeLeo VA (2005) Etiological factor in skin cancers: environmental and biological. In: Rigel DS et al. (eds) Cancer of the Skin. Elsevier Saunders, Philadelphia, PA, pp 61–70

Schlegel K (2006) Space Weather and Alexander von Humboldt's Kosmos Space Weather 4 doi:10 1029/2005SW000166

Schonberg M, Chandrasekhar S (1942) On the evolution of the Main Sequence stars. Astrophys J 96:161–73

Schopf JW (1999) Cradle of Life. Princeton University Press, Princeton, NJ

Selleri F (ed) (1998) Open Questions in Relativistic Physics. Apeiron, Montreal

Shapiro R (1986) Origins. Heinemann, London

Sharifi MH, Arozullah M (1986) Comparison of satellite and fiber optics technologies for intercity and intercontinental communications. Proc Int Conf Comm, Toronto, 2:827–831

Sher L (2001) Genetic studies of Seasonal Affective Disorder and seasonality. Comp Psych 42:105–110

Shirley JH, Fairbridge RW (eds) (1997) Encyclopedia of Planetary Science. Kluwer, Dordrecht

Silverman SM (1992) Secular variation of the aurora for the past 500 years. Rev Geophys 30:333–351

Singer SF (1993) Ozone depletion theory. Science 261:1101–1102

Smith RC et al. (1992) Ozone depletion: ultraviolet radiation and phytoplankton biology in Antarctic waters. Science 255:952–959

Stuiver M, Becker B (1993) High-precision calibration of the radiocarbon time scale AD 1950–6000 BC. Radiocarbon 35:35–65

Stuiver M et al. (1998) High-precision radiocarbon age calibration for terrestrial and marine samples. Radiocarbon 40:1127–1151

Suess HE (1980) The radiocarbon record in tree-rings of the last 8000 years. Radiocarbon 22:200–209

Summerer L (2006) Solar power satellites – European approach. Proc Jap Solar Power Conf, Kobe 2003

Swartz TE et al. (2007) Blue-light-activated histidine kinases: two-component sensors in bacteria. Science 317:1090–1093

Taylor HR et al. (1988) Effect of ultraviolet radiation on cataract formation. New Eng J Med 319:1429–1433

Tinetti G et al. (2007) Water vapour in the atmosphere of a transiting extrasolar planet. Nature 448:169–171

Tinsley B (1988) The solar cycle and the QBO influences on the latitude of storm tracks in the North Atlantic. Geophys Res Lett 15:409–412

Trieb F, Müller-Steinhagen H (2007) Europe-Middle East-North Africa cooperation for sustainable electricity and water. Sustainable Sci. 2:205–219.

Tsujimoto M et al. (2002) X-ray properties of young stellar objects OMC-2 and OMC-3 from the Chandra X-ray Observatory. Astrophys J 566:974–981

Tudge C (2005) The secret life of trees. Allen Lane, Harmondsworth

US Department of Energy (2000) Passive Solar Design Office of Building Technology. Oak Ridge National Laboratory, Oak Ridge, TN

US Department of Energy (2006) Solar Energy Technologies Program. Multi-Year Program Plan 2007–2011. DOE: Washington, DC

Vallina SM, Simó R (2007) Strong relationship between DMS and the solar radiation dose over the global surface ocean. Science 315:506–508

van der Spek B (2007) www livius org/cg-cm, 27 Oct 2007

van Dover CL et al. (1996) Light at deep-sea hydrothermal vents. Geophys Res Lett 23:2049–2052

van Dulken S (2002) Inventing the 20th Century. British Library, London

van Helden A (1989) Galileo, telescopic astronomy, and the Copernican system. In: Taton R, Wilson C (eds) Planetary Astronomy from the Renaissance to the Rise of Astrophysics Part A: Tycho Brahe to Newton. Cambridge University Press, Cambridge, pp 81–105

Vázquez M, Hanslmeier A (2006) Ultraviolet Radiation in the Solar System. Springer, New York

Vita-Finzi C (2002) Monitoring the Earth. Terra, Harpenden

Vita-Finzi C (2008) Fluvial solar signals. Geol Soc Spec Pub 296:105–115

Wallén CC (1955) Some characteristics of precipitation in Mexico. Geogr Annlr 37:52–58

Webber WR, Higbie PR (2003) Production of cosmogenic Be nuclei in the Earth's atmosphere by cosmic rays: its dependence on solar modulation and the interstellar cosmic ray spectrum. J Geophys Res 108, doi:10.1029/2003JA009863

Wharton JR, Cockerell CJ (1998) The Sun: a friend and enemy. Clin Derm 16:415–419

Wheeler PE (1994) The thermoregulatory advantages of heat storage and shade-seeking behavior to hominids foraging in equatorial savannah environments. J Hum Evol 26:339–350

WHO (1994) Ultraviolet Radiation. World Health Organization, Geneva

WHO (2001) Blindness: a global priority for the twenty-first century. Bull World Health Org 79. World Health Organization, Geneva

Wikipedia (2008) Northeast Blackout of 2003. en.wikipedia.org/wiki/2003_North_America_blackout

Williams GE (1990) Precambrian cyclic rhythmites: solar-climatic or tidal signatures? Phil Trans Roy Soc Lond A 330:445–457

Willman AJ Jr (2007) Known planetary systems. http://www princeton edu/~willman/planetary_ systems

Willson RC, Mordvinov AV (2003) Secular total solar irradiance trend during solar cycles 21–23. Geophys Res Lett 30:3,1–4

Wolstencroft RD, Raven JA (2002) Photosynthesis – likelihood of occurrence and possibility of detection on Earth-like planets. Icarus 157:535–548

Wright S (2007) Space debris. Phys Today 60:35–40

Zahradka K et al. (2006) Reassembly of shattered chromosomes in Deinococcus radiodurans. Nature 443:569–573

Zeuner FE (1959) The Pleistocene Period. Hutchinson, London

Zhong D-P (2007) Ultrafast catalytic processes in enzymes. Curr Opin Biol 11:174–181

Zirin H (1992) Breaking out in spots. Nature 359:271

Index

A

Abri Blanchard, 3
Abri Pataud, 1
ACE, 21, 119
Acne, 96
Adenosine triphosphate (ATP), 82, 121
Aeschylus, 122
Air travel, vii
Alaska, 1, 114
Albedo, 62, 64, 68, 92, 112
Alexander, 8
Alpher, Ralph A., 28
Alvin, 79
Amenhotep, 7
Amino acids, 81, 84
Ammonia (NH_3), 63, 81
Anaxagoras, 25, 84
Ångström, 13
Antarctica, 23, 80
Antikythera instrument, 3
Antioxidant, 99
Antrobus, viii
Apollo, 17, 45, 46, 111, 115, 116
Aristarchus, 10
Aristotle, 31, 65
Asimov, I., 1, 122
Asteroids, 25, 85
Astrology, 10
Astronaut, 43, 45, 92, 108, 111, 115, 116, 119
Astronomical unit (AU), 15, 19, 21, 33, 34,
    41–43, 65, 117
Atahualpa, 7
Atmosphere, terrestrial, 14, 15, 17–19, 22, 31,
    33, 34, 44, 46, 47, 52, 53, 61–64, 73,
    74, 79, 82, 91–93, 95, 111, 115, 119,
    121, 122, 134; early, 63, 64, 82, 85;
    lunar, 37; solar, 27, 28, 32–34;
    on HD189733b, 42

Aurelian, 7
Aurora, 16, 33–35, 47, 51, 53–55, 107–109,
    111, 113, 114
Australia, 19, 28, 58, 67, 75, 83, 86, 98, 101,
    104–106, 119, 126
Australopithecus, 89
Aztec, 7, 8

B

Bab al Mandab, 87
Babcock, H. W., 34, 49, 51
Babylon, 8, 10
Bach, Johann Sebastian, 58, 59
Bacon, Roger, 4
Balloon, 15, 17, 73, 94
Bang, Big, 28, 32
Becquerel, Alexandre-Edmond, 127
Bell, Alexander Graham, 109
Bering Strait, 75
Beryllium 10 (^{10}Be), 22, 24, 47, 68
Bethe, Hans, 28
Bible, 26, 44
Biomass, 92, 121, 133
Bipedalism, 89
BISON (Birmingham solar oscillation
    network), 19
Blackout, 54, 113, 114, 119
Bletchley Park, 110
Bolometer, 15, 44
Bridgers, Frank, 123
British Antarctic Survey, 93
Bromofluorocarbon
    (BCFC), 93
Bruno, Giordano, 37, 38, 41, 42
Bubble sextant, 5
Burj al-Taqa, 123, 132
Butti, Ken, 122